Assessing Revolutionary and Insurgent Strategies

CONCEPTUAL TYPOLOGY OF RESISTANCE

Paul J. Tompkins Jr., USASOC Project Lead

Jonathon B. Cosgrove and Erin N. Hahn, Authors

United States Army Special Operations Command
and
The Johns Hopkins University Applied Physics Laboratory
National Security Analysis Department

ASSESSING REVOLUTIONARY AND INSURGENT STRATEGIES

The Assessing Revolutionary and Insurgent Strategies (ARIS) series consists of a set of case studies and research conducted for the US Army Special Operations Command by the National Security Analysis Department of the Johns Hopkins University Applied Physics Laboratory.

The purpose of the ARIS series is to produce a collection of academically rigorous yet operationally relevant research materials to develop and illustrate a common understanding of insurgency and revolution. This research, intended to form a bedrock body of knowledge for members of the Special Forces, will allow users to distill vast amounts of material from a wide array of campaigns and extract relevant lessons, thereby enabling the development of future doctrine, professional education, and training.

From its inception, ARIS has been focused on exploring historical and current revolutions and insurgencies for the purpose of identifying emerging trends in operational designs and patterns. ARIS encompasses research and studies on the general characteristics of revolutionary movements and insurgencies and examines unique adaptations by specific organizations or groups to overcome various environmental and contextual challenges.

The ARIS series follows in the tradition of research conducted by the Special Operations Research Office (SORO) of American University in the 1950s and 1960s, by adding new research to that body of work and in several instances releasing updated editions of original SORO studies.

VOLUMES IN THE ARIS SERIES

Casebook on Insurgency and Revolutionary Warfare, Volume I: 1927–1962 (Rev. Ed.)
Casebook on Insurgency and Revolutionary Warfare, Volume II: 1962–2009
Case Studies in Insurgency and Revolutionary Warfare: Algeria 1954–1962 (pub. 1963)
Case Studies in Insurgency and Revolutionary Warfare—Colombia (1964–2009)
Case Studies in Insurgency and Revolutionary Warfare: Cuba 1953–1959 (pub. 1963)
Case Study in Guerrilla War: Greece During World War II (pub. 1961)
Case Studies in Insurgency and Revolutionary Warfare: Guatemala 1944–1954 (pub. 1964)
Case Studies in Insurgency and Revolutionary Warfare—Palestine Series
Case Studies in Insurgency and Revolutionary Warfare—Sri Lanka (1976–2009)
Unconventional Warfare Case Study: The Relationship between Iran and Lebanese Hizbollah
Unconventional Warfare Case Study: The Rhodesian Insurgency and the Role of External Support: 1961–1979
Human Factors Considerations of Undergrounds in Insurgencies (2nd Ed.)
Irregular Warfare Annotated Bibliography
Legal Implications of the Status of Persons in Resistance
Narratives and Competing Messages
Special Topics in Irregular Warfare: Understanding Resistance
Threshold of Violence
Undergrounds in Insurgent, Revolutionary, and Resistance Warfare (2nd Ed.)

SORO STUDIES

Case Studies in Insurgency and Revolutionary Warfare: Vietnam 1941–1954 (pub. 1964)

TABLE OF CONTENTS

LIST OF ILLUSTRATIONS

LIST OF TABLES

ACKNOWLEDGMENTS

The authors gratefully acknowledge the following authors and publishers for permitting the use of material in this report.

Figure 8. Types of counterinsurgency models. Reproduced with permission from RAND Corporation (Santa Monica, CA): Stephen Watts, Jason H. Campbell, Patrick B. Johnston, Sameer Lalwani, and Sarah H. Bana, *Countering Others' Insurgencies: Understanding U.S. Small-Footprint Interventions in Local Context* (Santa Monica, CA: RAND Corporation, 2014). © RAND Corporation.

Figure 10. Movement structures. Reproduced with permission from *ANSESRJ:* Jurgen Willems and Marc Jegers, "Social Movement Structures in Relation to Goals and Forms of Action: An Exploratory Model," *Canadian Journal of Nonprofit & Social Economy Research (ANSESRJ)* 3, no. 2 (Autumn 2012): 67–81.

James Jasper's strategic dilemmas pertaining to rationale are presented with permission from *Mobilization:* James M. Jasper, "A Strategic Approach to Collective Action: Looking for Agency in Social-Movement Choices." *Mobilization: An International Journal* 9, no. 1 (February 2004): 1–16.

Table 5 is reproduced with permission from *Perspectives on Politics*: Paul Staniland, "States, Insurgents, and Wartime Political Orders," *Perspectives on Politics* 10, no. 2 (June 2012): 243–264.

Content in Table 16, Types of nonviolent and rightful resistance tactics, published with permission from the Albert Einstein Institution. Originally published in Gene Sharp, *The Politics of Nonviolent Action* (Boston: Portor Sargent, 1973), and reproduced by the Albert Einstein Institution, "198 Methods of Nonviolent Action," http://www.aeinstein. org/wp-content/uploads/2013/09/198_methods-1.pdf.

INTRODUCTION

The Assessing Revolutionary and Insurgent Strategies (ARIS) project consists of a series of case studies and research conducted for the US Army Special Operations Command (USASOC) by the National Security Analysis Department (NSAD) of the Johns Hopkins University Applied Physics Laboratory (JHU/APL). Current and ongoing research efforts expand the ARIS mission into the development of analytical tools and methodologies to facilitate the deep and robust study of resistance, which is conceptually framed as the overarching phenomenon[a] that encompasses a broad spectrum of disruptive movement types, both violent and nonviolent.

Resistance is defined in this work as a form of contention or asymmetric conflict involving participants' limited or collective mobilization of subversive and/or disruptive efforts against an authority or structure. To better understand the fundamental attributes[b] of this phenomenon, the ARIS team developed this conceptual typology of resistance (hereinafter called "the typology" or "ARIS typology"). This effort seeks to both organize the interrelated concepts[c] essential to resistance in a formalized kind hierarchy and identify how these concepts are related to each other.

The typology is best described as an integrated kind hierarchy of individual concept typologies directly applicable to resistance groups and movements. The conceptual typologies are organized within the overarching attributes of resistance and their interior categories. As will be outlined in this work, the typology incorporates original, adaptive, and derivative work into a single construct, contributing to the research and literature as an instrument for concept development in the study of resistance. In this way, the typology provides a starting point for the formulation and evaluation of explanatory claims that can then be tested and verified through both qualitative and quantitative research methodologies.[1]

---

[a] The term *phenomenon* here refers to an event or series of events that can be observed and studied. In this typology, *phenomenon* is used only in reference to resistance—the complex phenomenon the authors seek to organize by identifying related attributes and concepts.

[b] The term *attribute* here refers to an inherent or fundamental characteristic, and *characteristic* refers more generally to a distinguishing trait or quality.

[c] The term *concept* here refers to an abstract idea or notion.

1

OBJECTIVE AND METHODOLOGY

The kind hierarchy of conceptual typologies[d] proposed here for the phenomenological study of resistance is part of the ongoing ARIS project. Since its inception, ARIS has explored past and current revolutions and insurgencies to identify emerging trends, an effort that has revealed two major takeaways:

1. Resistance, which encompasses a broad spectrum of disruptive movement types and manifestations, is an observable phenomenon with complex and dynamic characteristics and concepts.

2. The study of this phenomenon is spread across numerous disciplines, with little to no structure or common terminology through which one can apply research results outside his or her disciplinary context.[3]

These takeaways revealed a need in the research of resistance for the development of a typology of the phenomenon—a structured conceptual organization of the fundamental attributes of resistance, further detailing typologies of the related concepts within each attribute.

Conceptual typologies are descriptive, establishing space within which to characterize types that constitute "a kind of" in relation to an overarching concept and its defining attributes. In academic research, conceptual typologies are used for "rigorous concept formation and measurement" at the foundation of studies that are both methodologically robust and conceptually innovative.[4] The ARIS typology is best described as a hierarchical system of discrete concept typologies directly applicable to resistance groups and movements. The conceptual typologies are organized within the overarching attributes of resistance and their interior categories. This effort seeks to contribute original, adaptive, and derivative work to the research and literature as an instrument for concept development in the study of resistance. The ARIS typology provides a starting point for the formulation and evaluation of explanatory claims that can then be tested and verified through case study and other research methodologies.[5]

---

[d] Herein the word *typology* is used in reference to conceptual typologies, defined as "a form of typology that explicates the meaning of a concept by mapping out its dimensions, which correspond to the rows and columns in the typology. The cell types are defined by their position vis-à-vis the rows and columns. May also be called a descriptive typology."[2]

The methodological development of the conceptual typology of resistance included an interdisciplinary literature review (covering the fields of political science, sociology, economics, history, and law) and multiple collaborative analysis events with subject matter experts in December 2014 and February 2015. Subject matter experts included thought leaders in each of the disciplines addressed in the literature review, as well as numerous members of the USASOC community of interest. The events facilitated the review and discussion of issues concerning the study of resistance, eventually leading to the development of five primary attributes of resistance, each accompanied by numerous related concepts for further development. The team then adopted the collaborative research results as the structural and ideational foundation on which to build a deeper, more comprehensive, and rigorous typological construct through both original work and the integration or adaptation of typologies proposed in the literature by numerous scholars.

Table 1. Template for two-dimensional conceptual typologies.

	Variable 1	
	Value 1a	Value 1b
Variable 2		
Value 2a	Type A	Type C
Value 2b	Type B	Type D

Note: The title of each table is used to present the overarching concept of each typology.

Whenever possible, this effort uses the basic template for conceptual typologies and categorical variables shown in Table 1; this template was outlined by David Collier, Jody LaPorte, and Jason Seawright in both *The Oxford Handbook of Political Methodology* and their 2012 *Political Research Quarterly* article, "Putting Typologies to Work: Concept Formation, Measurement, and Analytic Rigor."[6] Two-dimensional typologies first depict the overarching concept being described, presenting two or more variables and their potential values. The cross-tabulation of component variables and their values then creates a matrix, and these values are then characterized as conceptual types in relation to the overarching concept.[7] Although primarily two-dimensional typologies are presented in this multifaceted and multilayered attempt to conceptualize resistance typologically, binary, unidimensional, and three-dimensional typologies are also presented.

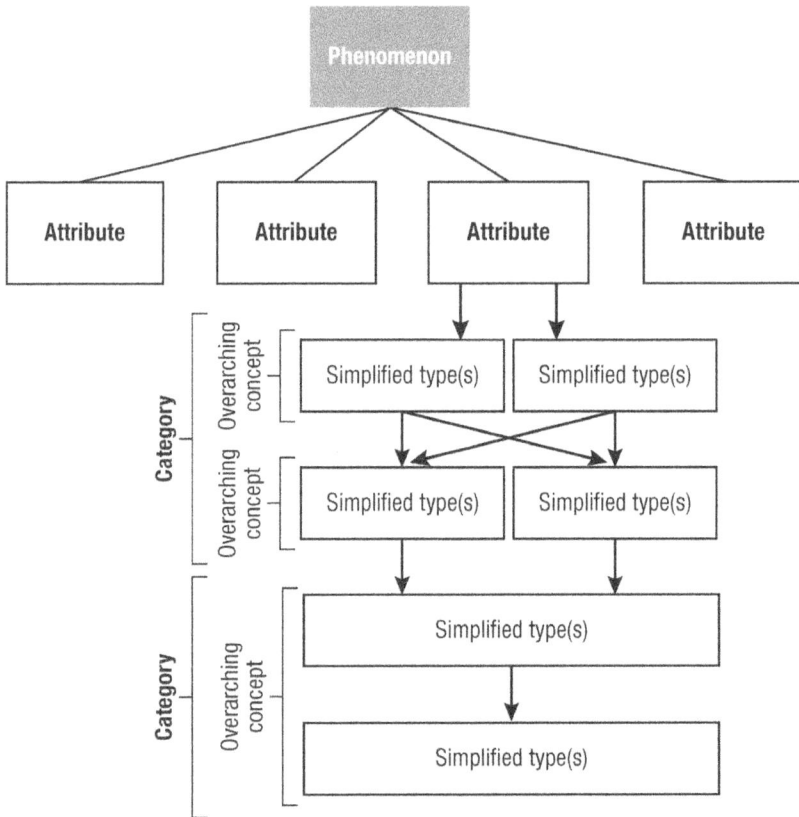

Figure 1. Template for typological kind hierarchies.

Each individual typology is organized within a five-tiered kind hierarchy (see Figure 1), which describes the "ordered relationship among concepts, in which subordinate concepts may be understood as 'a kind of' in relation to superordinate concepts."[8] The first level is the phenomenon itself (resistance). Second, the phenomenon is broken down into its fundamental attributes. Third, each attribute is divided into two or more categories. Fourth, discrete conceptual typologies are then presented within their applicable categories, indicated by their overarching concept and divided into two or more distinct types. Finally, in some cases there is a fifth level of typological subcategories within each distinct type. To begin each relevant section of this paper, kind hierarchies present the typological structure within each attribute. It is important to note that the graphical representation of each kind hierarchy contains only simplified portrayals of individual typologies and indicates ideational context rather than an exhaustive or formal development of each concept.

This comprehensive conceptual typology of resistance enables the formulation of rigorous questions for comparative research within a common framework, allowing for the cross-examination and applicability of research results. When applied in concert with an improved theoretical phasing construct of resistance (an ARIS task in parallel development), research based on this typology can then be further refined for operational applicability based on the movement's state of development. In this way, the typology of resistance can be the foundation on which a methodologically robust science of resistance can be built. For USASOC, this science of resistance would constitute the bedrock of scholarly knowledge informing its key mission areas—a robust and organized field of study answering key strategic questions of relevance to the operator, which are then considered and filtered for contribution to the development of training, doctrine, and depth of strategic thinking.

ATTRIBUTES

The attributes of resistance are the essential components of the phenomenon, arrived at through the course of two collective analysis meetings (December 2–3, 2014, and February 11–12, 2015) of interdisciplinary subject matter experts. This process resulted in five distinct yet interrelated attributes (see Figure 2).

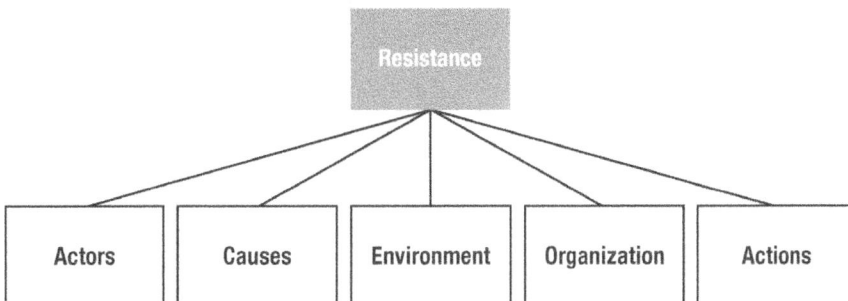

Figure 2. The attributes of resistance.

1. **Actors:** The individual and potential participants in an organized resistance, as well as external contributors and either competing or cooperating resistance groups

2. **Causes:** The collectively expressed rationales for resistance and the individual motivations for participation

3. **Environment:** The preexisting and emerging conditions within the political, social, physical, or interpersonal contexts that enable or constrain the mobilization of resistance, directly or indirectly

4. **Organization:** "The internal characteristics of a movement: its membership, policies, structures, and culture"[9]

5. **Actions:** The means by which actors carry out resistance as they engage in behaviors and activities in opposition to a resisted structure; can encompass both the specific tactics used by a resistance movement and the broader characteristics or repertoires for action (i.e., strategy)

Each attribute contains a series of relevant typologies (one-, two-, and three-dimensional), defined by their key characteristics. This paper presents a high-level view of the typological content of each attribute before presenting each typology in detail. Many scholars and researchers who have engaged the topic of resistance have proposed useful typological contributions to the body of knowledge, many of which are presented here either in whole or in an adapted form. This typology aims to be as thorough as possible while simultaneously restricting its scope to resistance as a phenomenon of human behavior (meaning the individuals, organizations, and movements), rather than encompassing the factors that have impacted or can impact cases of resistance across the disciplines ad infinitum.

WHERE RESISTANCE BELONGS IN THE IDEATIONAL SPACE

A preceding ARIS manuscript supporting this effort, titled *Developing a Typology of Resistance: A Structure to Understand the Phenomenon*, presented a multidisciplinary review of how the phenomenon of resistance is diversely studied, demonstrated the academic and operational need for a conceptual typology of resistance, and established an ideational foundation from which to move forward in development of a typology.[10] Essential characteristics of resistance include participants' foundational use of agency against an opposed structure, the asymmetry (at least initially) of the resistance group or movement relative to the opposition, the largely contentious nature of confrontations, and the subversive nature and tactics used by the group, movement, and participants.

Resistance is a form of conflict involving the collective and subversive efforts of participants against an authority or structure. Broad in conceptual scope, but also limited in reach, resistance can be carried out through either violent or nonviolent means (or both) on either an international or an intranational scale. While this conceptualization is deceptively far-reaching in nature, resistance is particularly concerned with participants' collective, asymmetric action against a relatively well-established opponent, excluding much of traditional political conflict as a whole (i.e., interstate conflict and disputes). Resistance need not be political; some resistance movements are focused on purely social, religious, economic, or ethnic factors, or on a complicated layering of two or more such focuses (see "Structural Focuses of Resistance Rationales").

ACTORS

The actors attribute broadly encompasses the individual and potential participants in an organized resistance, as well as external contributors and either competing or cooperating resistance groups. Direct actors are those who will inevitably emerge as actors within any resistance group or movement, namely the leadership and participants thereof. Other actors are classified as indirect or potential. This categorization includes members of the general population who may be loyal to or sympathize with the resistance, other resistance groups operating in the same operational or rhetorical space, and potential external supporters of the resistance (see Figure 3).

Direct actors are those individuals active within the resistance group or movement itself, while indirect or potential actors are those who are either peripheral to the conflict (general population) or may or may not enter into the resistance scenario or efforts (other resistance groups and external support). Conceptually, any resistance group or movement must have leaders and participants, no matter how small in number or informal in execution. Indirect or potential actors, on the other hand, either may not leverage their influence (e.g., the public may remain indifferent, or external support may not materialize) or may not exist (e.g., there may be no other active resistance groups).

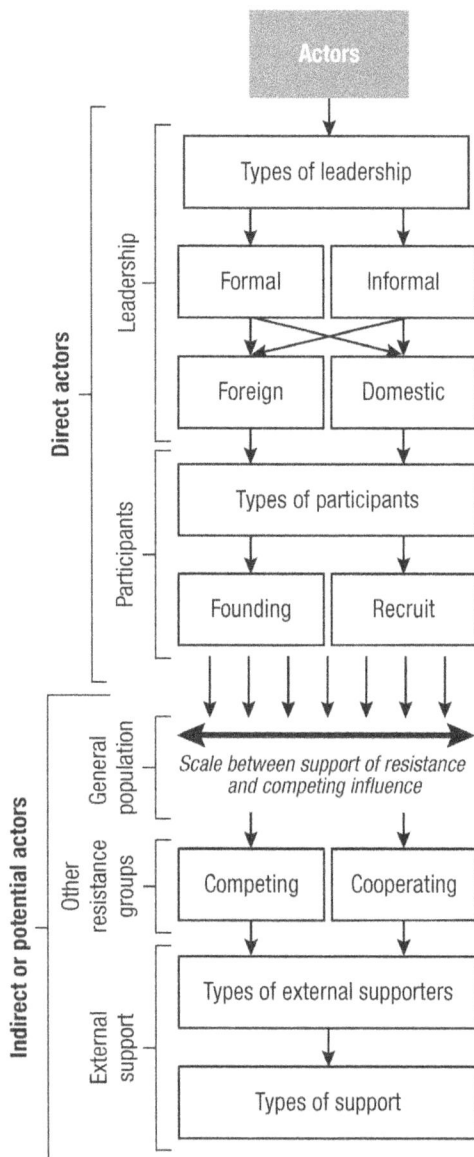

Figure 3. Actors kind hierarchy.

Leadership

The leaders of resistance movements are those individuals within a movement or organization who "provide strategic and tactical direction, organization, and the ideology of the movement," performing "these functions within the unique and compelling context of their country, culture, and political economy."[11] In both violent and nonviolent resistance, leadership roles and functions are extremely diverse across roles, functional areas, styles, and other characteristics.

Types of Leadership

The types of dominant leadership in a given resistance movement or organization can be characterized according to style, function, or both. It should be noted that stylistic and functional typologies of resistance leadership overlap in some respects but are both nevertheless valuable in their own right and merit inclusion in the typology for potential use in the study of resistance.

A stylistic typology of resistance leadership is offered by Rex D. Hopper in his seminal article, "The Revolutionary Process: A Frame of Reference for the Study of Revolutionary Movements," presenting the leadership roles of agitator, prophet, reformer, statesman, and administrator-executive. First, the agitator is usually informal, being either one who "stirs the people not by what he does, but by what he says . . . [leading] people to challenge and question . . . [the status quo and] create unrest" or one who serves "to intensify, release, and direct tensions which people already have."[12]

Second, a prophet is a leader who "feels set apart or called to leadership" and claims "special and separate knowledge of the causes of unrest and discontent," speaking "with an air of authority . . . in general terms." Usually instrumental in the formulation and promulgation of "the social myth" of the resistance, a prophet leader uses "his belief in himself and his confidence in his message as a means of articulating the hopes and wishes of the people."[13] Third, a reformer is one who "attacks specific evils and develops a clearly defined program," attempting "to change conditions in conformity with his own conceptions of what is good and desirable."[14]

Fourth, a statesman is one who is "able to formulate policies and will attempt to carry social policy into practice" and who "will propose the program which promises to resolve the issues and realize the objectives

9

of which the people have become aware."[15] Finally, an administrator-executive is one who fully implements the policies of the movement, completing the formal institutionalization of the goals of the resistance. Such a leader is likely to present dynamics and characteristics of the other four stylistic types but is distinguished from the rest through the concrete implementation of policies.[16]

Beyond the more broad-stroke stylistic typology of resistance leadership as outlined by Hopper, the type of leadership can also be characterized according to more discrete categories of functional role, expertise, or particular authority within the movement or organization. Composed of those categories produced by both ARIS collaborative analysis event participants and other sources, these functional leader types include political, military, ideological, religious, intellectual, economic, diplomatic, financial, administrative, scientific, technical, and professional leaders.[17]

Leadership Characteristics

The leadership of a resistance group or movement can be typified according to several other characteristics. First, a resistance leader or cadre of leaders can be geographically foreign or domestic. Top leaders for some movements, particularly those that are violent and seek broad international impact, may choose to direct resistance efforts from a remote or foreign location. Conversely, domestic leaders conduct their business within the country or region where resistance activities are taking place.

Second, the leadership of a resistance group or movement can be characteristically formal or informal. Formal leaders or leadership cadres are those that are either established or assigned to their posts by superior authorities in the group or movement or chosen by the group through a predetermined and legitimized process. Informal leaders, on the other hand, naturally emerge to or seize the position through charisma, personal ambition, or magnetic dynamism in the form of a cult of personality. Most of the types of leaders delineated above typically fall into the category of formal (e.g., statesman, administrator-executive) or informal (e.g., agitator, prophet), but this is not an ironclad principle, and deviations from these norms would prove to be analytically significant.

Participants

The participants in the resistance movement or organization can be sorted within a two-dimensional typology, dependent on their role and level of participation (see Table 2). The first significant variable in discerning types is identifying a participant's role as either martial (i.e., militaristic, meaning an at least partially formalized armed role within the organization or movement) or nonmartial. Martial roles are characterized by an armed and violent participation in resistance, particularly within a stronger and more formal hierarchical command and control structure. While all martial roles are inherently violent in nature, not all nonmartial roles are nonviolent; nonmartial participants can serve as violent agitators or provocateurs. Participation levels are divided between high, meaning a deep, nearly full-time dedication to the resistance effort, and low, meaning a superficial but recurring part-time participation.

Table 2. Types of participants.

	Role of Participant	
	Martial	**Nonmartial**
Level of Participation		
High (deep; full-time)	Fighters	Core membership/ underground
Low (superficial; part-time)	Sleeper cell	Supporter/auxiliary

Note: Core membership and supporter roles/functions are included to ensure that the category of participants is not exclusive to armed resistance, allowing for its use in studying nonviolent resistance movements.

Those participants with a high participation level in nonmartial roles in the movement or organization can be considered among the core membership or underground. Those serving nonmartial roles at a low level of participation can be considered supporters or auxiliary. On the other hand, those who participate at a high level in a martial role are fighters, variously referred to as guerrillas, insurgents, or other paramilitaries within the movement or organization. Similarly, those participants who serve a martial role at a low participation level can be considered violent actors in sleeper cells.

Information technology's rapid emergence and the Internet's wide proliferation have far-reaching implications on many aspects of human society, including warfare and resistance. Cyberspace has been recognized as a new domain in the practice of warfare,[18] and the United States and other nations (as well as resistance groups) continue to develop both defensive and offensive cyber capabilities. Although there are well-established perceptions of what military operations look like in the land, air, sea and space domains, activities in cyberspace are so young that no clear differentiation has yet been made between military cyber warfare and other nonmartial, subversive cyber actions. However, research and legal necessity will increase demand for this distinction to become a major point of research in the coming years.

Core Membership and Underground

This type of participant is primarily dedicated to the resistance, devoting the majority of his or her time to its pursuit and activities. Core membership can be defined as those participants integrated into at least one facet of the resistance organization's operations on a full-time basis. In violent resistance, the core membership participants manifest as an underground because they must operate secretly. The underground is "a clandestine organization established to operate in areas denied to armed or public components [of a resistance] or conduct operations not suitable for the armed or public components."[19]

The functions performed by the core nonmilitant membership of resistance movements are numerous. Some of these functions may be shared or supported by auxiliary participants.[c] Functions typically unique to core membership and undergrounds include strategic planning, finances, security, and various operational functions including subversion, psychological operations, sabotage, shadow government management, and selective recruitment for leadership, intelligence, and other special tasks. Psychological operations may include public relations, broad communication, mass response, and violent coercion. Certain types of particularly sensitive intelligence activities (scene-of-battle, sabotage, scientific, military, and political) will also typically fall to core membership in the underground.[20] Likewise, human resource

---

[c] Some functions may also be rendered unnecessary in the case of nonviolent resistances.

12

management, basic administrative tasks, and research and development tasks are functions normally reserved for core membership.

Supporters and Auxiliary

Supporters can be defined as those participants who are actively dedicated to supporting the resistance but do so only on a part-time basis on the periphery of their work and lifestyle, contributing their efforts when needed. In the context of violent resistance and parallel to an underground, the auxiliary are

> the support element of the [resistance] organization whose organization and operations are clandestine in nature and whose members do not openly indicate their sympathy or involvement with the [resistance] movement. Members of the auxiliary are more likely to be occasional participants of the insurgency with other full-time occupations.[21]

Auxiliary participants often share or support some functional roles largely pursued by the underground component. These functions overlap with some filled by the underground and core membership and include mass recruitment, communications (as couriers and messengers), logistics, storage, procurement, labor for material fabrication, transportation, intelligence collection, propaganda distribution, early warning security, safe house management, and medical/social services.[22]

Fighters

The fighters type of participant can be defined as those "organized along military lines to conduct military and paramilitary operations"[23] in a resistance, but this type also includes those similarly organized for other forms of "subversion and violence."[24] Although underground and core members may be used for similar violent and subversive functions, the distinct organization of fighters according to military principles makes them typologically different.

Sleeper Cells

The category of sleeper cells emerged from the typological research as a potential type in the two-dimensional relationship between participant roles and participation level. Theoretically, these participants would be organizationally managed in a military fashion but would remain dormant within a given population until called on to conduct guerrilla or terrorist operations.

Participant Characteristics

While the individual participants' motivations to join the resistance group or movement are covered in the "Causes" section, another typologically significant aspect of individual participants is the differentiation between founding members and recruits. A founding member is one who was a participant in the resistance group in its earliest stages. Recruits, however, are those who were incorporated into the group or movement as participants once the resistance was already under way and seeking to grow its ranks. This typological distinction in research and analysis of historical cases could yield numerous operationally vital insights into movement dynamics.

Researching questions concerning whether participants are founding or recruited members could potentially lead to significant insights on emerging resistance movements and their viability for growth or success. Such questions could include the average proportion between founding members and recruits under certain operational or security environments, the tendency of certain types of movements or groups to have more or fewer founding participants on average, and the average recruitment rate of violent as opposed to nonviolent resistance movements. These factors could in turn be examined in relation to the movement's relative success or failure in achieving objectives or perpetuating its influence.

General Population

Support from the general population (or mass base) is an important factor in any type of resistance movement and a vital asset to both resistance organizations and those loyal to the resisted structure or government. This typology is expressed on three one-dimensional scales (see Figure 4), each traveling in different directions away from indifference or fence-sitting (when an individual holds no preferential sentiment or loyalty toward either faction):

1. Support for the resistance movement or organization (left)

2. Support for the resisted structure, organization, or government (right)

3. Support for other parallel resistance groups or movements that may exist (center)

The scale or commitment of support for any of these competing interests ranges in descending value from committed support as a participant in resistance or loyalist to the resisted authority, passive support, sympathy, and bystander interest in either cause.

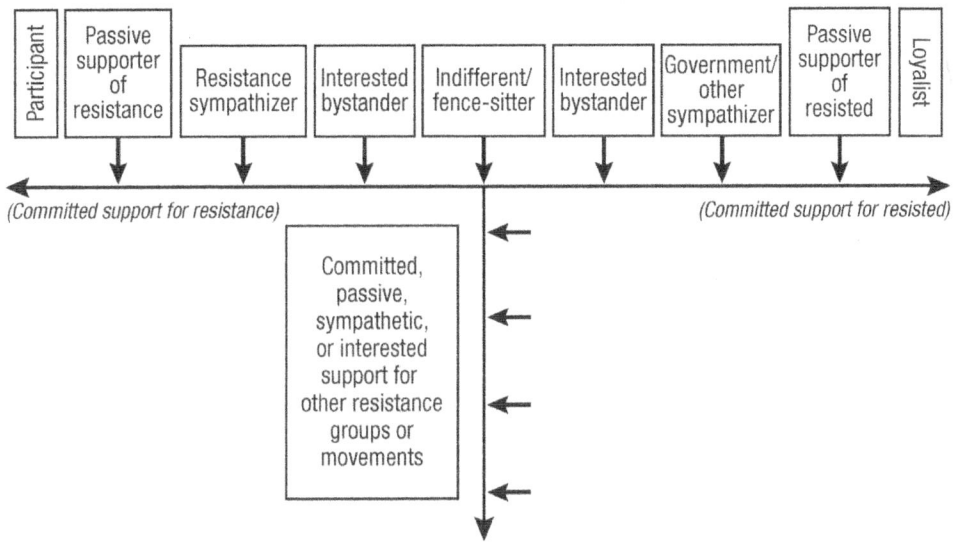

Figure 4. General population.

Ideally, appropriate data for the weighting or statistical distribution of popular support among the general population would be collected through polling data precise enough for this five-valued scale. If existing data is not precise enough for this typology, one can use a simplified typology of popular support. One example would be the "range of popular support" presented in US Army doctrine on *Tactics in Counterinsurgency* (FM 3-24.2), which presents three more generalized values of popular support to a government or insurgency that are applicable to resistance as a whole: indifference, passive support, and active support (see Figure 5).[25]

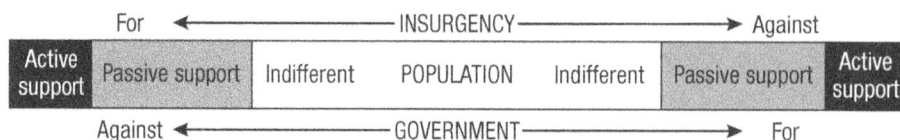

For	◄——————————— INSURGENCY ——————————►	Against

| Active support | Passive support | Indifferent | POPULATION | Indifferent | Passive support | Active support |

Against	◄——————————— GOVERNMENT ——————————►	For

Figure 5. Range of popular support presented in FM 3-24.2.

Other Resistance Organizations

As stated above, the context of any given resistance is not a vacuum, and other resistance groups or organizations may exist at the same time and in the same geographic space, ideational space, or both.

Typologically, when considering or studying a single resistance group or movement, one can categorize other organizations within the same national or regional context as competing, cooperating, or unrelated resistance organizations. Competing groups may be either counterresistance movements or members of the same resistance movement in contention over ideational goals, strategic issues, or limited resources. Additionally, the range of competition between groups can vary from secretive tensions to vocal opposition to open and violent hostilities.

Cooperating resistance organizations may or may not be members of the same overall movement and may cooperate for a range of practical, tacit, ideological, or strategic reasons. Merely expedient cooperation may prevail when the groups' goals intersect over the short term, reverting to competition and back to cooperation as circumstances change. Finally, unrelated resistance organizations are only unrelated in their (possibly intentional or temporary) lack of a positive or negative relationship with the group or movement in question. The typological title of unrelated does not signify that the two movements have no overlap whatsoever but merely that a cooperative or competing relationship with the given resistance is not emergent at that time.

External Support

The provision of support by entities or individuals external to the country can be a decisive factor for the longevity and potential victory of a resistance group or movement. For this reason, obtaining effective and consistent external support is often a high priority for modern resistance movements. The types of potential external supporters include foreign governments, nongovernmental organizations, transnational criminal groups, other non-state actors (similar resistance groups, private interests, etc.), diaspora from abroad, and foreign fighters who travel to support the conflict.

Table 3. Types of external support.

	Form of Support	
	Nonmaterial	Material
Assertiveness of Support		
Passive	Moral	Sanctuary
Active	Political	Resources

How external actors support a resistance can be effectively portrayed with a two-dimensional typology, with cell types derived from US Army doctrine (see Table 3).[26] When the overarching concept is external support, one can organize the four types according to the form of the support (nonmaterial or material) and the assertiveness of that support (passive or active). First, moral support constitutes a nonmaterial form of passive or tacit support, only going so far as sympathetic public statements or similar measures. Second, political support is active and nonmaterial, including advocacy and symbolic actions to express committed support. Third, sanctuary constitutes a material form of passive support through the provision of secure training sites, operational bases, protection from extradition, or other shields from adversary actions through the inaction of the external supporter (e.g., failure to extradite is not activity but a failure or refusal to act). Finally, the provision of resources (funds, weapons, food, advisers, training, foreign fighters, etc.) is active material assistance, the most involved type of resistance support that can be offered by an external actor.

CAUSES

Simply put, this attribute encompasses the collectively expressed rationales for resistance and the individual motivations for participation. The rationale for resistance has two primary components: first, the structural focus (i.e., the portion or portions of society in which the movement or group seeks to enact change); and second, the type of rationale, characterized by what change is sought as well as how radically and on what scale it is sought. Individual motivations, however, include the motivations for initial participation, those for continued participation, and those for exiting the movement or organization (see Figure 6).

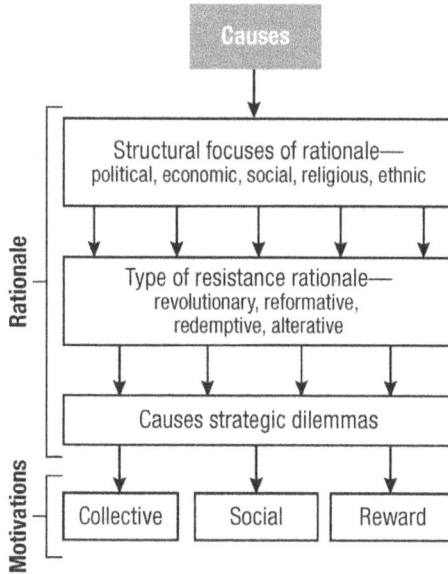

Figure 6. Causes kind hierarchy.

It is important to note that in this typology, the word *causes* is not used in reference to theoretical causality but instead in reference to, as defined in *Merriam-Webster's Collegiate Dictionary*, "something (such as an organization, belief, idea, or goal) that a group or people support or fight for" or "a reason for doing or feeling something."[27] Conceptually, this attribute is related to narrative frames in the study of social movements and contentious politics. This concept was framed as "orientation" by Patrick Regan and "aims and motives" by Bruce Hoffman.[28] Clarification of causes as a concept distinct from causality is best presented by Daniel Byman, who posited that insurgents need "a cause, and with it a narrative and associated propaganda." Byman continues, "rebels fight for, or at least against, something . . . [and] often turn this cause into a broader narrative that includes a critique of the current order and plans for how the rebels would rule should they come to power."[29] In this way, the causes attribute is concerned with the expressed goals and fixation of the resistance movement and participants. Causal relationships between variables, on the other hand, should be developed through careful research and analysis.

Rationale

The rationale of a given resistance movement or organization is the collectively or organizationally propagated narrative for collective action, outlining the resistance movement's values, claims, and objectives. A group uses a narrative to delegitimize the resisted authority, legitimize its own claims, garner the general population's sympathy or support, and encourage recruitment. In other words, an organization or movement uses a narrative to make its case as to why its cause is right and that of its adversary is wrong.

The rationale serves as the overall structure to provide what social movement theory characterizes as narrative frames for mobilization of collective action among groups. As described by Robert D. Benford and David A. Snow, the narrative structures of social movements are divided into diagnostic, prognostic, and motivational frames. Diagnostic frames describe the problem and identify victims, prognostic frames articulate a proposed solution and strategy, and motivational frames provide the population with the reason for engaging in collective action, including the development of a suitable vocabulary to mobilize individuals.[30] The rationale propagated by a resistance group is effectively the whole, with each of the three narrative frames forming the essential components.

Structural Focuses of Resistance Rationales

Resistance movements develop rationales that seek to credit their actions as proscriptive to solve one or more structural issues in society. These issues can be political, economic, social, religious, and/or ethnic. While formalized here, this reality is reflected in the common descriptions of resistance movements in academic, military, and press analyses when one seeks to typify a particular movement (e.g., is the Islamic State primarily a religious, political, or social movement, or a weighted combination of all three?). In this way, one can describe a resistance movement as sociopolitical, ethnosectarian, socioeconomic, or politicized socioreligious, or by evoking numerous other societal focuses.

By identifying the particular spheres of life and society a resistance movement seeks to impact, an analyst can refine the selection of cases in order to isolate and consider variations in movements. Including this additional typological layer when studying the internal workings of otherwise typologically identical resistance rationales allows the analyst to examine potential deviations in the data. This rigor is enabled by an explicit statement of research assumptions and limitations on the data. For example, "This research will assume that the slightly divergent weighting of resistance rationale priorities between socioeconomic movements X and Y are not a significant factor in relation to the dependent variable and therefore can both be considered types in kind."

Resistance Movement Rationale

The typology of rationales for resistance is two-tiered in nature, first differentiating between four macro-types, each of which contains numerous discrete types (see Table 4). The first level is an adapted form of David F. Aberle's seminal classification of social movements as presented in his book *The Peyote Religion among the Navaho*, which differentiates movements along two dimensions: "the locus [or scope] of the change sought" and "the amount [or extent] of change sought."[31]

Table 4. Types of resistance movement rationales

	Extent of Impact Sought	
	Limited/Specified	Radical/Total
Scope of Impact Sought		
Sweeping (societal, national)	Reformative	Revolutionary
Specified (communities, individuals)	Alterative	Redemptive

Adapted from David F. Aberle's typology of social movements.[32]

Aberle's construct defined the locus values as "individual" and "supra-individual."[33] However, it is important to note that within the context of resistance movements (as opposed to social movements as a whole), even those movements that seek a limited or specified impact are less concerned with enacting change in individual circumstances than with achieving their goals or vision at a local or community scale. For this reason, the values for scope have been adapted to be characterized as either sweeping (societal, national) or specified (communities, individuals).

The resulting macro-types of resistance movements according to rationale are revolutionary, redemptive, reformative, and alterative movements (each defined below). Distributed between these four macro-level typological categories of resistance rationale are several further distinguished micro-level types. While the macro level is defined by the scope and extent of change sought by the movement, the micro-level types are discerned according to what kind of change is sought in relation to the existing status quo. The majority of these subtypes were derived and adapted from the nine "types of insurgency" developed by Bard E. O'Neill in his book *Insurgency & Terrorism: From Revolution to Apocalypse*, with limited contributions from elsewhere in the literature.[31]

The hierarchical juxtaposition of O'Neill's types within the larger framework of Aberle's typology demonstrates within the literature a multidimensional approach to examining resistance movements' rationales and objectives.

Revolutionary Movement Rationales

Revolutionary movements seek to overthrow the standing system to which they are subject. Depending on the particular movement's objectives, it may seek to establish a new structure in place of the old. This is often characterized as a transformation of society, wholly or partially. There are five general subtypes of revolutionary movement rationales: pluralist, egalitarian, traditionalist, anarchist, and apocalyptic-utopian.

First, pluralist revolutions seek to destroy or displace the existing system or status quo in favor of a new structure beneficial to participants, usually rationalized as promoting individual freedoms. Second, egalitarian revolutions are those that wish to destroy the existing system or structure to impose one founded on equal distribution of social and/or physical goods (often through centralized control) and the mobilization of groups to radically transform community structures to this end. Third, traditionalist revolutions seek to replace a given system or structure with one based on the application of articulated "primordial and sacred values," often rooted in ancestral or religious lineage. Fourth, anarchist revolutions strive for the permanent destruction of a given institutionalized system or structure, discarding all authority patterns as illegitimate and unnecessary. Finally, apocalyptic-utopian revolutions (which are usually partially religious in nature) are those that either "envisage establishing a world order . . . as the result of an apocalypse" or seek to destroy the current order in preparation for such an eschatological event.[35]

Reformative Movement Rationales

Reformative movements seek a specific change or changes to the existing structure or system on a large scale, often characterizing the targeted change as key to improving society as a whole. The three general types of reformative rationales are reformist, preservationist, and restorationist.

First, reformist resistance movements are those that seek either autonomy within the current system or relatively modest change to the status quo. Second, preservationist reformative movements are those that seek change to maintain the existing system or structure. Such movements may engage in actions against nonruling groups or movements or those authorities who are acting to disrupt the status quo.[36] Finally, restorationist movements seek "to restore an elite group opposed to an occupying authority [or any other authority to recently gain influence] in order to regain power."[37]

Redemptive Movement Rationales

Redemptive movements seek dramatic change among some individuals or a specified community, often in the form of a complete transformation of the specified person(s) or their circumstance. In the context of resistance, two types of redemptive movements can take shape: secessionist or fundamentalist. First, secessionist or separatist resistance movements are those that seek to withdraw from the status quo system or structure (rather than destroy it) to establish a new, independent system or structure.[38] Political secessionist movements in particular are dynamically different from other resistance movements because they are not center-oriented toward power or influence held by their opponents. Second, fundamentalist resistance movements seek the significant transformation of community or individual behavior and self-governance according to stringent ideological principles, usually as separate from the rest of society.

Alterative Movement Rationales

Alterative movements seek limited or specified change among some individuals or a particular community, often concerning the way people think about certain behaviors or issues within a given system or structure. In the context of resistance, at least three alterative movements can be observed: migratory, expressivist, and commercialist. First, migratory resistance movements are those in which participants make or seek a physical or associational move from one system, structure, or country to another. Second, expressivist resistance movements are those that wish to modify individuals' reactions to and thinking about unpleasant external realities that participants feel powerless to change.[39] Finally, commercialist resistance movements are those that are driven primarily by participants' acquisition of material resources and wealth through the very act of resistance or the acquisition of power.[40]

Strategic Dilemmas Pertaining to Rationale

As presented in James M. Jasper's article "A Strategic Approach to Collective Action: Looking for Agency in Social-Movement Choices" (and presented here with permission), studying explicit choices and implicit trade-offs resistance organizers and participants face opens a "fruitful new path of research" that can "represent agency in contrast to the structure that has interested scholars for so long."[41] Typologically, a resistance group or movement confronts strategic dilemmas of rationale, organization, action, and information. Each type of dilemma is presented in the section describing its respective attribute (information dilemmas are presented with actions). In research, strategic decisions may present both compelling dependent and independent variables, asking either how the strategic decisions of the group or movement impact later development or how characteristics of the resistance or the environment may impact strategic decisions.

The rationale for resistance presents at least two strategic dilemmas for the group or movement. First, the shifting goals dilemma asks whether the resistance should "stick with [its] original goals" and try "to find the right means" to achieve them or whether the goals should be adjusted to fit the present "abilities and opportunities." Although the former option might make the movement more likely to succeed, it may also alienate participants and undermine the legitimacy of the movement.[42] Second is the dilemma of inevitability, in which "an ideology that suggests you must eventually win offers confidence but makes collective action less critical,"[43] potentially undermining recruitment and participation.

Motivations

The motivations for resistance are the factors that drive individuals to participate in a resistance, manifesting "as a function of the perceived attractiveness or aversiveness of the expected consequences [costs and benefits] of participation."[44] There are three overarching types of motivation for participation in a resistance movement, originally developed by Bert Klandermans in his *American Sociological Review* article "Mobilization and Participation: Social-Psychological Expansions of Resource Mobilization Theory."[45] Each type of motivation is theoretically in play and "no one factor is preeminent," collectively forming a "multiplicity of motives" in the participant, each of which is valued more or valued less relative to the other motives.[46]

First, the collective motive is a function of the personal "expectation that participation will help to produce [a] collective good and the value of [said] collective good."[47] The ARIS publication *Human Factors Considerations of Undergrounds in Insurgencies* frames the collective motivation as "belief in the cause or political factors" as ideology can play a role in both recruitment and retention.[48] In this way, the collective motive constitutes the only type of individual motivation directly related to the collective rationale of the resistance. The remaining motives that incentivize (or disincentivize) participation in resistance are at most indirectly related to the resistance rationale.

Second, social motives emerge from the costs and benefits of participation "as distinguished in the reactions of significant others."[49] Such motives can include group norms (including "what their friends and comrades think of them"), prestige or recognition (when a group seeks to retain participants, this is accomplished through "morale-sustaining techniques"), and the avoidance of social alienation or condemnation. The simple inertia of mobilization and habit toward resistance can also constitute a motive for joining, or staying with, a resistance group or movement.[50]

Finally, the reward motive includes the "non-social costs and benefits" that result from participation in collective action toward resistance.[51] Most numerous, and likely most influential, among the three motive types, the reward motive can include numerous "personal and situational factors" related to the individual participant's "problems and . . . immediate needs," which are leveraged by the resistance for recruitment through "propaganda and promises" of both reward and the removal of "government persecution." However, these positive incentives can be paired with measures of coercion against the individual, constituting a promise of nonsocial costs associated with refusal to participate.[52] Once a participant has already joined the resistance, the group may use surveillance and threats of retaliation to keep the participant within the ranks of the group.[53]

ENVIRONMENT

The broadest attribute of resistance, the environment encompasses the preexisting and emerging conditions within the political, social, physical, or interpersonal contexts that may enable or constrain the mobilization of resistance, directly or indirectly. In other words, the environment is everything outside of the resistance that can impact or shape it (see Figure 7). The consideration and study of environmental factors is integral to the study of any social phenomenon, so this section reviews several factors and characteristics believed to be formative or impactful on resistance according to existing research literature. However, this typology seeks to conceptualize the components of "resistance" for future systematic study by testing hypotheses against both component and environmental factors. Because most environmental factors are conceptually separate and distinct from resistance itself (e.g., although regime type may impact the formation or practice of a resistance, it is not a component of the resistance itself), the environmental factors reviewed here are not developed into formal typologies. The only exception to this rule are environmental factors or characteristics that exist because of the presence of resistance (e.g., counterinsurgency efforts by a government), thus making them part of the phenomenon itself.

Notable Characteristics

Potentially preexisting or emerging factors external to resistance are nearly infinite, requiring the researcher to seek out and investigate those factors that may significantly affect the shape or outcome of resistance movements, or those environmental outcomes of resistance movements (e.g., failed states, authoritarian regimes, democratic institutions) that are strategically significant. During an ARIS collaborative analysis event (February 2015), participants offered numerous notable factors and characteristics of the environment that merit special consideration in the analysis of resistance. Although typological elucidation of the majority of these environmental characteristics falls outside the scope of a typology of resistance, these factors nevertheless stand as some of the most compelling variables for consideration in the comparative study of resistance movements and provide an opportunity for introducing an interdisciplinary scope to the investigation of this phenomenon.

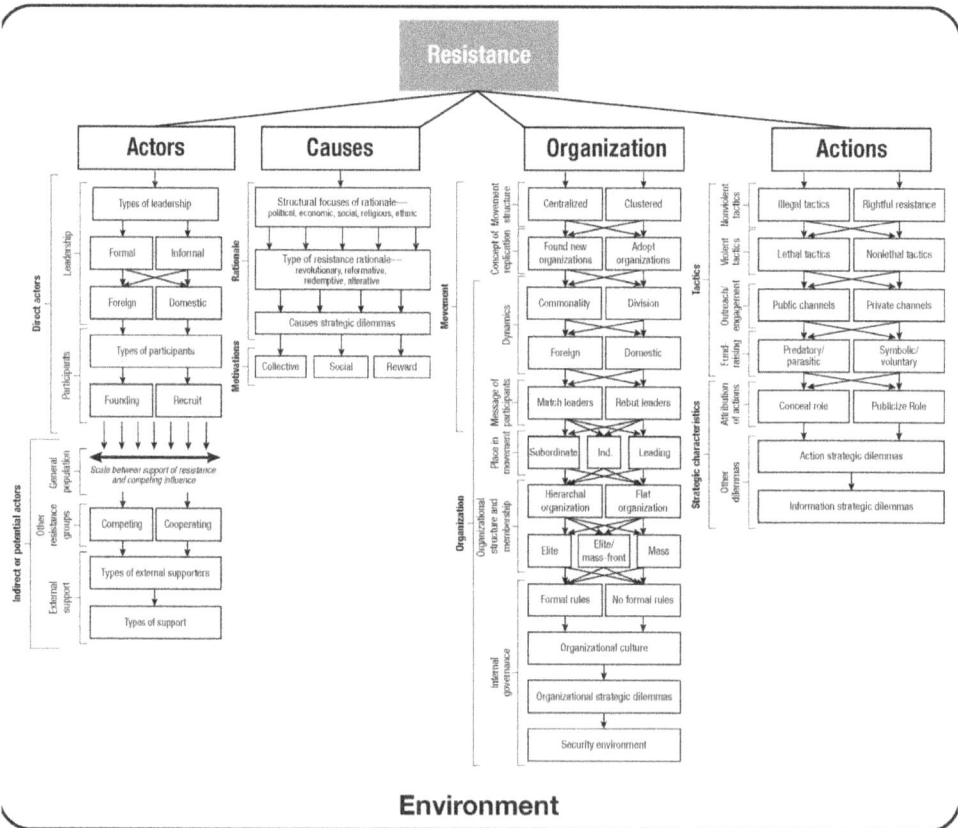

Figure 7. Environment relative to kind hierarchies.

Environmental factors and characteristics that merit special consideration in the formation of research questions when examining the phenomenon of resistance include those listed in the following subsections.

Characteristics of the State

Particularly in relation to political resistance movements and any resistance that uses illegal and/or violent tactics, the characteristics of the state are extremely relevant environmental factors for analysis. First, some characteristics may serve to fuel or defuse a resistance movement's rationale. Such factors as state rigidity, the distribution of power, elections, popular access to institutions, and the ability to meet the population's needs may be considered. Second, some state characteristics may be instrumental in determining the failure, success, or stalemate of a political resistance against political authorities. Examples of these include military and law enforcement capacity and government structure.

Social Structures

The dominant patterns and customs of the country, region, and participants can also play a key role in the shape and development of resistance movements, whether that movement is at least partially political or concerned with other issues (economic, social, religious, ethnic). Cultural, ethical, religious, familial, ethnic, gender-based, and many other accepted societal patterns can be compelling attributes of violent and nonviolent resistance movements for the purpose of analysis. How old or entrenched is the norm, and how did it impact perception of or participation in the resistance? How do resistance movements address societal norms—do they reject them as counter to the cause or appeal to them for mass support? Is there a relationship between these approaches and various types of resistance rationales?

Law

The legal contexts (local, national, regional, international, religious, the philosophical basis of law, etc.) within which a resistance must operate can be a fundamental factor, particularly in the analysis of resistance strategy, tactics, and the barrier to collective action. While many means of nonviolent public activism are legal and protected in

Western and industrialized countries, the imposing legal structures of authoritarian and totalitarian countries can strictly outlaw even some forms of speech or association, a factor that could greatly impact participation, organization, strategy, and tactics.

Economy

The economic environment, both preexisting and resulting from resistance, can be examined as either potentially causal to resistance or resulting from the success, failure, or stalemate of certain types of resistance movements. Such factors include unemployment, gross domestic product, recession, social mobility, inflation, private ownership, and many others.

Technology

Popular access to, and the growing capabilities of, various technologies is a factor that can be examined as potentially directly formative on tactics and strategy, as well as indirectly on the shape or rationale of resistance movements.

Communications

Arguably the most central arena of resistance movements is that of ideas and their dissemination, making the popular modes of communication and Internet penetration a notable variable for analysis against the other attributes and characteristics of resistance. The means and instruments of communication available to a resistance movement can be particularly influential on a resistance movement's organization, coordination, growth, strategy, and tactics.

Education

The availability, quality, and average level of education among the general population can be examined as both a causal factor in various types of resistance (e.g., is there a relationship between the average level of education and participation in political versus social or religious resistance movements?) and a formative variable on the rationale, information, engagement, and recruitment strategies of the resistance.

Geography

The various types and details of geography (terrain, etc.) and climate, as well as the distribution and density of the human geography, can impact the formation and development of resistance movements.

Time

On multiple levels, time is a notable variable for the analysis of resistance. Important considerations include the longevity of different types of resistance movements (based on any one of their numerous attributes and characteristics), the ability of past events or injustices featured in resistance rationales to affect participation and sympathy among the general population, and the strategic and informational value of symbolic dates.

Relationships

Preexisting and emerging relationships among individuals, organizations, various social groups (social, class-based, religious, ethnic), and governments through personal, diplomatic, security, historical, or other channels and contexts can be significant in the analysis of the formation, shape, and success of resistance movements. A contemporary example of preexisting relationships shaping resistance movements is the role of Baathist military officers in the growth, organization, and rise of the Islamic State, which later became the most prolific insurgent terrorist group in Iraq, Syria, and elsewhere in the region.[54]

Many such relationships become central to resistance networks and dynamics (see "Movement Structures" and "The Dynamics of Resistance Movements or Organizations"), but some connections may emerge from the event of resistance itself. For this reason, preexisting relationships among resistance participants should be conceptualized as an environmental factor, while those concurrent with or emerging from resistance are more suitably considered an organization attribute. The relationships among the opponents of resistance, both preexisting and emerging, are conceptualized as environmental factors relative to the resistance movement.

Typologically Relevant Environmental Characteristics

Some environmental factors are particularly relevant to resistance movements, and thus more detailed typologies are presented here. These environmental conditions are directly related to resistance, in that they cannot occur outside the existence of a resistance movement.

Counterinsurgency Efforts

The operations and policies of a government entity "designed to simultaneously defeat and contain insurgency" and potentially "address its root causes"[55] are typologically relevant environmental factors that usually exist only when a resistance movement is developing or present. These operations and policies are intentionally targeted and meant to negatively impact resistance efforts. While many characteristics of the state (see above) undoubtedly impact the efficacy of any given type of counterinsurgency model, such factors (e.g., state resources, efficacy of law enforcement or military) can be accounted for among the research assumptions and limitations of a given historical case study.

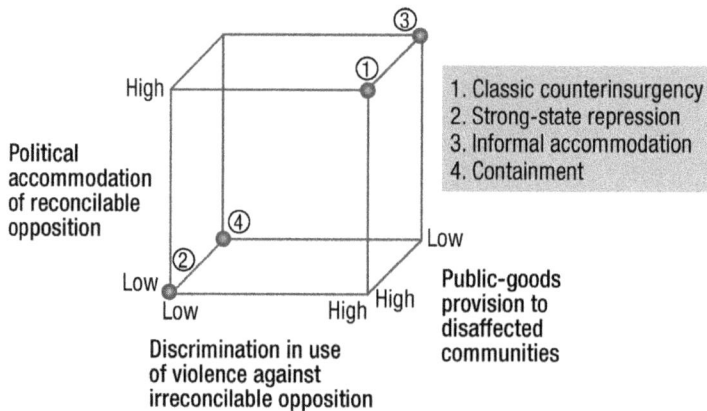

Figure 8. Types of counterinsurgency models.

A robust three-dimensional typology of counterinsurgency models (see Figure 8) was developed by Stephen Watts et al. in the RAND National Security Research Division study *Countering Others' Insurgencies: Understanding U.S. Small-Footprint Interventions in Local Context.* This typology, presented here with permission, represents an exemplary breakdown of the categories that can be deduced regarding models for counterinsurgency. It is important to note that these are four "ideal types," and many counterinsurgency models will differ in some respects.[56]

In a three-dimensional structure, the typology of counterinsurgency models outlines types based on their combination of either high or low emphasis on three principles: political accommodation of reconcilable opposition, discrimination in the use of violence against irreconcilable opposition, and public-goods provision to disaffected communities. Although the intersection of factors allows for eight possible typological combinations, only the four most noted permutations are developed. First, a classic counterinsurgency model emphasizes high values in all three principles (i.e., striving for accommodation, limited or carefully targeted use of violence, and the provision of public goods) and is "most often associated with . . . highly capable Western democracies."[57]

Second, although the strong-state repression counterinsurgency model emphasizes the provision of public goods to assuage discontent, the state is nevertheless "unable or unwilling to offer accommodations to any part of the opposition, relying instead on broad, often indiscriminate use of force and terror to suppress organized dissent" (e.g., Russia in Chechnya or the Soviet Union in Afghanistan).[58]

Third, an informal accommodation counterinsurgency model is apparent when the state is unable or unwilling to provide public goods to discontented communities, although "this does not mean that the regime cannot reach an accommodation with the armed opposition." Accommodation could range from the opposition's incorporation into the regime to the state's tacit allowance of the opposition's territorial control, a dynamic that also results in the selective use of violence against irreconcilable opposition.[59]

Finally, a containment counterinsurgency model describes situations in which a state is "unable or unwilling to accommodate the reconcilable opposition" and "uses force—usually indiscriminately—to repress insurgent activity," accepting risk of residual violence as long as the resistance is relegated to certain, often poorly defined, bounds. Regimes that leverage the containment model are also usually content with "concomitant limits on its control over portions of its population and territory," resulting in little to no concern for the provision of public goods to disaffected populations.[60]

Wartime Political Orders

In cases of violent resistance resulting in a state of civil war, various types of wartime political orders can emerge in the relationship between the distribution of contested control and the level of cooperation between the state and the resistance. The typological model of these power relationships, presented with permission in Table 5, were developed by Paul Staniland in his 2012 *Perspectives in Politics* article, "States, Insurgents, and Wartime Political Orders."[61]

Table 5. Types of wartime political orders.

	State–Resistance Cooperation		
	Active	Passive	Nonexistent
Distribution of Control			
Segmented	Shared sovereignty	Spheres of influence	Clashing monopolies
Fragmented	Collusion	Tacit coexistence	Guerrilla disorder

A segmented distribution of territorial control is one "in which each side controls some territory," whereas fragmented distribution of control constitutes those in which "both sides have presence throughout the area under contestation." The level of cooperation between the state and the resistance signifies the "dynamics of cooperation and bargaining" that operates "alongside violence and conflict" in civil war contexts.[62] When the state and resistance "actively cooperate towards a shared goal," this can result in a situation of shared sovereignty or collusion. Shared sovereignty "is a negotiated form of political order" in which "the state has not shattered its foe but instead the two sides have arranged a clear division of influence and authority that satisfies both in the pursuit of mutual gains."[63] Collusion, on the other hand, "is a situation in which the state actively cooperates with non-state armed actors that are geographically intermeshed with its areas of operation" in the "coordinated pursuit of a shared goal."[64]

The state and resistance can reach a level of passive cooperation through "live-and-let-live bargains and tacit deals that create implicit but often incomplete and tenuous arrangements for the management of violence." Spheres of influence emerge from arrangements "to limit the boundary violations against each sphere . . . [and] where state and [resistance] forces will tread," intending "to manage spirals of escalation."[65] Tacit coexistence, however, is characterized by the "fragmented, overlapping control" in "careful attempts to limit the degree of active conflict and violence between states and non-state armed groups in intermixed daily life."[66] Finally, the context of no cooperation "along any dimension" between the violent resistance and the state is when one observes total civil war, in which "unpredictable violence and unclear lines of authority and control characterize the interactions between fighters." The absence of cooperation in a segmented context causes clashing monopolies, where each side controls "distinct territory" and "the boundaries between state and non-state forces are rigid and easy to identify." Guerrilla disorder, however, "is a situation of fluid violence in which there are few clear norms or rules about the infliction of lethal violence when insurgent and state forces are intertwined in the same physical spaces" and "violence is an embedded part of political, economic, and social life."[67]

Environment as an Independent Variable

This attribute represents a rich source of independent variables, many of which are most intriguing in reference to macro-level questions of causality behind resistance movements. By including environmental characteristics as independent variables in research questions, analysts can pursue the variables' impact on the success or failure of the resistance itself (i.e., X variable is an enabler of resistance, or Y variable is a constraint to resistance). They can also address potential relationships between the characteristics of an emerging resistance and the persistent environmental conditions (i.e., there is a significant relationship between X environmental characteristic and Y organizational characteristic).

Examples of compelling research questions derived from this typology that use environmental factors as independent variables include the following:

- Is there a significant relationship between the presence of regional traditions of informal community leadership and the emergence of informal leadership as dominant in resistance movements?

- Is there a significant relationship between the established presence of authoritarian systems of government and the development of reformative rationales in resistance movements?

- Is there a significant relationship between high Internet penetration and the interorganizational structure of resistance movements?

- Is there a significant relationship between the use of lethal violent tactics in resistance movements and the presence of legal frameworks that highly restrict nonviolent means of political engagement and expression?

Environment as a Dependent Variable

Emerging environmental characteristics can also be examined as variables dependent on resistance characteristics. Significant historical and political science scholarship has examined the relationship between revolutionary movements and the emergence of certain types of political contexts (e.g., the use of terror by the revolutionary government, concentration of power, single-party system). However, there remains a persistent need for more precise strategic analysis of the relationships among the attributes of resistance movements and the emerging environmental factors resulting from their success, stalemate, or failure.

The strategically targeted broadening of formal, rigorous knowledge and the study of these relationships will allow for informed strategies and planning in both counterinsurgency and unconventional warfare operations. In the context of counterinsurgency planning, for instance, formal comparative studies of the relationship between sectarian-nationalist political resistance movements (e.g., the Muslim Brotherhood) and the types of regimes that usually emerge would provide a level of strategic awareness of the potential outcomes of such movements (e.g., the Arab Spring). Likewise, in unconventional warfare, the knowledge of the relationship between certain types of resistance movements and the regimes or outcomes that result could inform policy making and planning to determine which resistance movements should (or should not) be supported to coerce, disrupt, or overthrow a regime.

Examples of compelling research questions derived from this typology that use environmental factors as dependent variables include the following:

- Is there a significant relationship between the success of resistance movements that have charismatic informal leadership and the emergence of highly restrictive legal systems?

- Is there a significant relationship between the failure of ethno-religious preservationist resistance movements and the secularization of theocratic regimes?

- Is there a significant relationship between the repression of resistance movements with centrifugal or polycentric dynamics and the persistence of cadre relationships into future resistance efforts?

- Is there a significant relationship between the stalemate of a resistance movement that uses lethal violent tactics and the sharp decline in national gross domestic product?

ORGANIZATION

The organization attribute concerns "the internal characteristics of a movement [or group]: its membership, policies, structures, and culture."[68] A key distinction is required here between a resistance organization and a resistance movement. Resistance organizations are the individual groups acting for the cause, whereas the resistance movement constitutes the collective effort of one or more resistance organizations allied or coordinated toward similar or identical objectives. Although a resistance movement may be represented by only one organization, several typologies apply specifically to those that are composed of multiple groups. Likewise, several typologies apply to both movements and organization, and others apply only to the resistance organization (see Figure 9).

The Movement

Several types and dynamics are particular to the multiorganization resistance movement. Many of these have been the subject of extensive typological study in the past and have been adapted for presentation here.

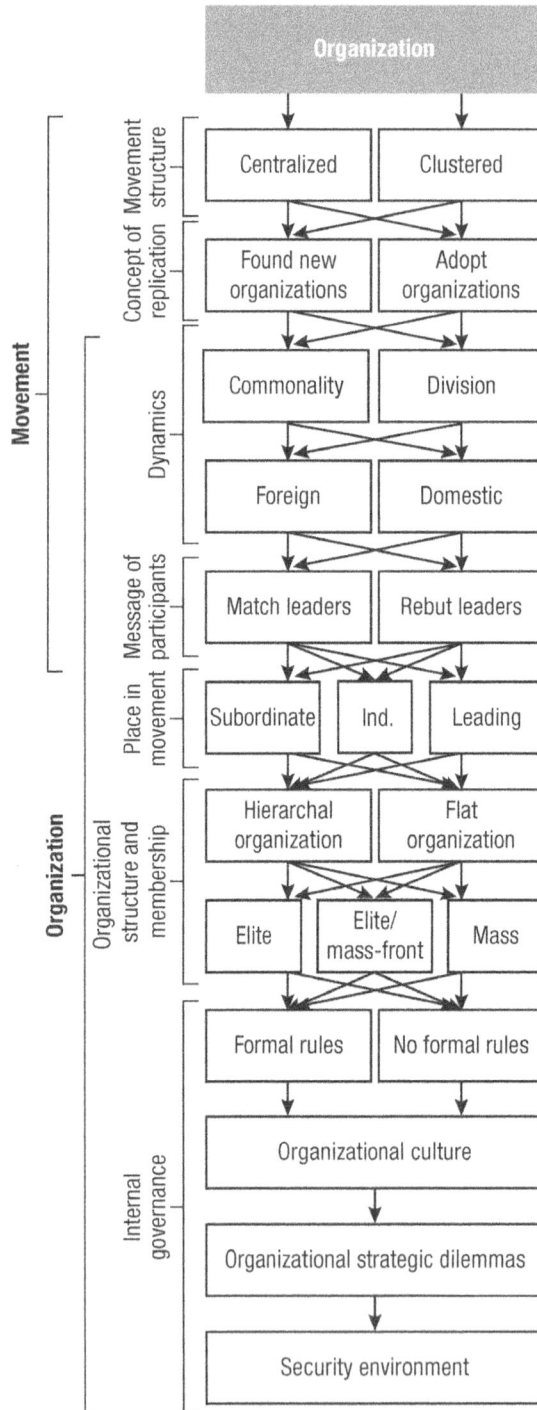

Figure 9. Organization kind hierarchy.

Movement Structures

The structure of a resistance movement can be defined "as the system of relations that exist between all entities involved in a given [resistance] movement" (or part thereof). This structure is defined by the degree of formality (i.e., the strength of network ties between the organizational nodes of the movement)[f] and the centralization of organized efforts under a single group or collective entity (i.e., hierarchical or lateral). Adapted from the typology proposed by Jurgen Willems and Marc Jegers[69] and presented here with permission, these two factors can be portrayed in a two-dimensional typology, where resistance movement structures can be categorized as formal-centralized, informal-centralized, formal-clustered, and informal-clustered relationships between resistance organizations (see Table 6 and Figure 10).

Table 6. Types of movement structures between organizations.

	Formality of Organizational Relationships	
	Formal	**Informal**
Authority Relationships between Organizations		
Hierarchical	Formal-centralized	Informal-centralized
Lateral	Formal-clustered	Informal-clustered

Hierarchical organizational relationships are characterized by one actor having authority or a decisive role over the other, which is derived from the actor's "ability and/or willingness to keep or share information [or other resources] with another actor."[70] Alternatively, lateral relationships demonstrate relatively equal power status and reciprocity between organizations, "based on mutual exchange of information [or resources]."[71] Regarding formality, informal relationships are characterized by their flexible "implicit and unwritten" nature, are primarily "trust based," and are derived from "culture, habits, and beliefs."[72]

---

[f] Several organization typologies concern networks, which can be analyzed and defined according to their nodes (individuals or organizations) and the ties between nodes. Ties can be strong or weak and vary by direction, density (number of ties), transitivity (influence of ties), connectivity (shortest paths between nodes), and betweenness (number of shortest paths through a particular node). Although thorough analysis and literature exist on these issues, the ARIS typology is concerned with conceptualization on the macro-scale and will use this terminology only for clarification purposes.

Formal organizational relationships, on the other hand, are "external-ized and/or recorded," reducing "uncertainty in the future" through a more "rigid" structure "based on a legal system and/or a set of widely accepted rules."[73]

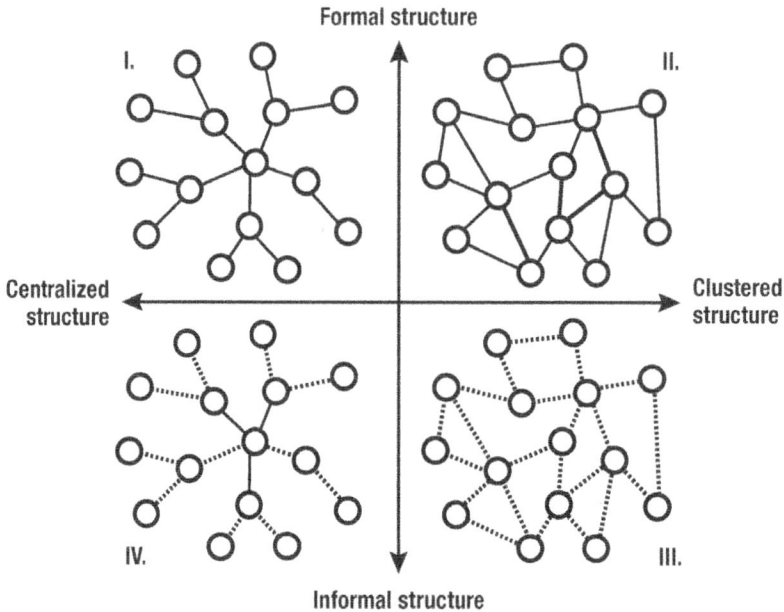

Figure 10. Movement structures.

Concept of Replication

Many resistance movements adopt a concept of replication to facili-tate a strategy for growth. Whether this is organized around the foun-dation of affiliated parties or groups or the adoption of existing groups into the movement, the concept of replication is characterized by the movement's growth outside the organization itself. Recruitment to the organization itself does not constitute a concept of replication. One can think typologically of resistance movements that do or do not adopt a concept of replication, and these concepts can include either or both the foundation of new movement organizations and the adoption of existing organizations into the resistance movement.

The Organization

Numerous dynamics and characteristics of the resistance organiza-tion can be shared with and typologically applied to the movement as a whole. However, values of some attributes apply to a specific organization and should be considered only within that organizational context (e.g., organization-specific typologies used in the study of the Muslim Brotherhood in Egypt would exclude sister movement organizations elsewhere in Africa or the Middle East, despite their being organizations within the broader Society of the Muslim Brothers). Each typology will specify whether it applies to both resistance organizations and movements, or to only organizations, and *group* can be considered synonymous with the word *organization*.

Dynamics and Characteristics

The Dynamics of Resistance Movements or Organizations

Resistance movements and organizations are never completely united in all things, and although participants share some sources of commonality, they inevitably have sources of division that come into play. The dynamic interaction between the sources of commonality and division can also be thought of as centrifugal and centripetal dynamics (i.e., those forces that drive participants apart or draw them together). The typology of these movement or organization dynamics is two-dimensional, defined by the combinations of strong or weak sources of division and commonality (see Table 7), which can emerge among or between leaders and participants from ideological, religious, social, ethnic, strategic, personal, or other sources. The terminology for this typology is largely derived from the SPIN (segmentary, polycentric, and integrated network) categorization of American social movements developed by Luther P. Gerlach and Virginia H. Hine.[74]

Table 7. Types of dynamics in resistance movements or organizations.

	Sources of Division	
	Strong	**Weak**
Sources of Commonality		
Strong	Polycentric dynamics	Cohesive dynamics
Weak	Segmentary dynamics	Networked dynamics

Movements or organizations with strong sources of commonality and weak sources of division among participants may be characterized as cohesive, possessing largely centripetal dynamics that reinforce unity of collective effort and identity. Likewise, those with strong sources of division and weak sources of commonality are segmentary ("composed of many diverse groups, which grow and die, divide and fuse, proliferate and contract") and characterized by centrifugal dynamics that drive participants apart and toward potential fissures. Those movements or organizations that exhibit strong sources of both commonality and division may be described as having polycentric dynamics ("having multiple, often temporary, and sometimes competing leaders or centers of influence"). Finally, those with weak sources of both commonality and division can be regarded as exhibiting networked dynamics ("forming a loose, reticulate, integrated network with multiple linkages"). Common manifestations of networked dynamics in resistance are alliances of convenience, in which there may be no strong division between the groups but their common enemy is one of only scarce reasons for them to cooperate. Networked dynamics can include "overlapping membership, joint activities, common reading matter, and shared ideas and opponents."[75] In other words, groups or movements with networked dynamics involve participants who have a high density of weak ties.

Foreign and Domestic Characteristics of Resistance Movements or Organizations

Any resistance group or movement will have foreign or domestic characteristics relative to its audience in the general population. These characteristics can be important in popular support, recruitment, strategy, tactics, and many other attributes of a resistance organization or movement. The typology of foreign and domestic characteristics of an organization or movement is two-dimensional, defined by the combination of the geographic base of operations and the popular perception of the group or movement (its leadership or participants) as foreign or domestic (see Table 8).

When a resistance organization or movement is perceived as domestic and operates primarily within the given country, it is a domestic resistance. Those organizations or movements perceived as foreign and that operate inside the country are transnational. It is important to note that a group or movement could conduct some transnational operations and activities but still be considered domestic in the relative context of the given country where it is most prevalent. However, those resistance groups or movements that are seen as foreign and operate primarily outside the given country are foreign (e.g., in the context of the United States, the Irish Republican Army is a foreign resistance group, as it is both rooted and active in foreign contexts). Finally, groups or movements that are perceived as domestic but primarily operate in foreign contexts are displaced, the most notable example being governments in exile and other resistance groups forced out of their domestic context by environmental factors (e.g., government oppression) or strategic choice.

Table 8. Types of foreign and domestic characteristics in resistance movements or organizations.

	Geographic Base of Operations	
	Operates Primarily Inside Given Country	Operates Primarily Outside Given Country
Perception of Group/Movement as Local or Foreign		
Perceived as foreign	Transnational	Foreign
Perceived as domestic	Domestic	Displaced

Messaging of Resistance Organizations or Movements

The strategic imperative of effective messaging and representation of the resistance movement or organization to various audiences can play a significant role in the apparent efficacy or appeal of resistance efforts. All movements seek to appear united, as the emergence of visibly centrifugal dynamics among participants may invite skepticism or reveal weakness, leading some resistance leaders to try and impose unity of message on the movement or group. A two-dimensional typology of unity of message in external representation is therefore shaped by whether the leadership seeks to impose the perception of a united front and whether the messaging matches or differs from that of the leadership (see Table 9).

When resistance participants mirror the narratives and objectives of a leadership that seeks to impose such a unity of message through either positive or coercive incentive systems, the external representation of the group or movement is a controlled message. However, when the leadership tries and fails to impose a unity of message, conflicting

41

messages emerge as participants contradict the line propagated by resistance authorities. However, some resistance movements and organizations may not seek to impose a united message, instead inviting diversity of opinion among participants. In this context, an open forum may take shape where participant's messages differ from that of leadership, or coordinated messages can emerge, either through naturally cohesive dynamics with messages derived from leadership or in a bottom-up, populist model that steers the messaging of the leadership based on that of participants.

Table 9. Types of unity of message for resistance movements.

	Messaging Goal of Leadership	
	Impose Unity of Message	**Invite Diversity of Message in Participants**
Message of Participants		
Matches leadership	Controlled messages	Coordinated messages
Differs from Leadership	Conflicting messages	Open forum

Organizational Position in Overall Movement

A resistance organization can hold numerous positions within the wider resistance movement, depending on the level of coordination within the movement and the subordinate, autonomous, or leading role played by the organization (see Table 10). Within collectively coordinated efforts, there are dependent organizations, autonomous members of coalitions, and head organizations. However, within separate and uncoordinated efforts, there are mother and spin-off organizations, as well as autonomous resistance groups.[76]

Table 10. Types of organizational positions in resistance movements.

	Organization Independence and Seniority		
	Subordinate	Autonomous	Leading Role
Coordination of Efforts between Resistance Organizations			
Collective effort	Dependent organization	Member of coalition	Head organization
Separate efforts (uncoordinated)	Client organization	Resistance organization	Mother organization

Organizational Theories and Membership

Theory of Organization

Resistance groups organize according to competing theories of how they should be structured, and many strategic, ideological, and other factors influence their choice of structure. These theories of resistance organization can be differentiated along a two-dimensional typology, based on the interaction between the concentration of organizational authority and the degree of organizational separation between the group leadership and the rank-and-file participants (see Table 11).

When authority is consolidated in a hierarchical structure, the result is vanguardism, in which a single leader (or small group) is the sole leadership and guardian of the cause. However, when consolidated authority exists in a flatter organizational structure, this is a mass line structure, in which key members of the leadership seek to reduce the participants' feelings of separation from the leader, likely casting him or her as "one of the people." When authority is shared or diffused at the top of a hierarchical structure, this can be characterized as decentralized or conciliar governance, in which numerous figures share authority, likely each over numerous branches of the organization. When this diffused authority exists in a flat organization, the theory of organization becomes consensual, with many decisions likely put to the group as a whole or heavily impacted by participants.

Table 11. Types of theories for authority in a resistance organization.

	Concentration of Authority	
	Consolidated	**Diffused or Shared**
Organizational Separation between the Leadership and Participants		
High (hierarchical)	Vanguardism	Conciliar
Low (flat)	Mass line	Consensual/democratic

Membership Strategies

Resistance organizations that seek internal growth must implement a strategy to incorporate recruits and new members. The core dynamics defining types of membership strategies are the barrier to membership and whether new participants are integrated into the core membership or used in front organizations. These characteristics create a two-dimensional typology of strategies that groups may use for security-related, strategic, and other reasons (see Table 12). The typology is primarily derived from *Undergrounds in Insurgent, Revolutionary, and Resistance Warfare*, originally written in 1963 and updated by the ARIS team in 2013.[77]

Exclusive or elite groups with a high barrier to membership (which can often incorporate tiered, trial-based membership processes) can either incorporate new participants into the core organization (elite organization) or use those new members in a front group outside the primary group (elite front organization). Inclusive groups with low barriers to membership, however, can either incorporate all new recruits into the mass organization or similarly incorporate the broad swath of recruits into front organizations despite the open membership approach (mass front organization). This conceptualization of ideal types, however, should not be construed to suggest that membership processes for most resistance organizations are overly formalized or orderly. In many movements, they are chaotic, ad hoc, and vary widely across the movement in absence of, or despite, a centralized strategy. Such inevitable complexities must not be ignored in the study of membership in a given movement.

Table 12. Types of membership strategies in resistance organizations.

	Barrier to Membership	
	High (Exclusive)	Low (Inclusive)
Integration of New Participants		
Integrated in base organization	Elite organization	Mass organization
Mostly leveraged through front organizations	Elite front organization	Mass front organization

Internal Governance

The internal governance of a resistance organization can take several forms, fitting within four general categories. First, there are formalized measures of discipline and rules, established by the group and leadership for the governance of participants. Second, the organizational culture constitutes the informal management of participants' behavior within the group. Third, decisions on various strategic dilemmas regarding the effective organization of the resistance group also contribute to internal governance. Finally, the security environment imposed on participants in the resistance governs the functions and communications among portions of the organization.

Discipline and Rules

Olivier Bangerter's research note *Regulating Armed Groups from Within: A Typology*, published by the Small Arms Survey in 2012, outlines eight types of internal discipline and rules a group may use for internal governance. This typology lists and defines several mechanisms, including oaths; codes of conduct ("the set of rules an organization expects its members to respect under all circumstances"); standing orders (which "differ from codes of conduct in that they define [behavior] that is expected in a specific situation as opposed to at all times. . . . Standing operating procedures—a subset of standing orders—spell out what a fighter or a unit must do when confronted with a given challenge"); operation orders; military manuals; internal organization documents (which "spell out the procedures to follow when taking decisions . . . address[ing] issues such as the command structure, the decision-making process, the responsibilities and powers associated with different positions . . . and how group members have to work together"); and penal or disciplinary codes.[78] Another instrument for internal governance are founding charters, which differ from internal organization documents because they more generally address the goals and vision of the resistance and are written for both internal and public audiences.

Organizational Culture

Resistance organizations, although unique from other organizations in many critical respects, can nevertheless be examined culturally along similar typological lines. Akin to other organizations, resistance groups are composed of individuals who must collectively order their efforts toward the achievement of collective goals. For this reason, a typology of cultures developed in the field of organizational studies and dependent on how members of the organization interact and relate to one another is compatible, despite not being developed with resistance groups in mind. However, there may be room for improvement or a need for the thorough development of an exclusive typology of resistance organizational cultures derived from the case literature.

In his 2002 book *Kultura Organizacyjna*, Dr. Czesław Sikorski presented a division of cultures according to "the attitude of the organisation members to cultural dissonance."[g] This concept was adapted by Łukasz Sułkowski of the University of Social Sciences in Poland (Społeczna Akademia Nauk) in his review of the literature, "Typologies of organisational culture—multi-dimentional [*sic*] classifications." The variable shown in the columns is the acceptance of cultural dissonance in the group, and the variable shown in the rows concerns the antagonism of interpersonal relationships between participants (see Table 13).[80]

---

[g] Sułkowski clarifies that "cultural dissonance is related to the existing differences between the ways [participants] think and behave in organisations, which are the most often reasons for conflicts and misunderstandings."[79]

Table 13. Types of organizational cultures (Sikorski and Sułkowski).

Interpersonal Relations	Acceptance of Cultural Dissonance	
	Acceptance	Lack of Acceptance
Antagonistic	Culture of rivalry	Culture of dominance
Nonantagonistic	Culture of cooperation	Culture of adaptation

First, an organizational culture of rivalry exists when cultural dissonance is accepted but interpersonal relations are antagonistic, resulting in a culture characterized "by strong rivalry" and in which "members believe that it is necessary to prove their superiority." Rivalry cultures can also be described as participatory, "collectivist," "heterogenic," "presence-oriented," and having "a clash of different cultural problems."[81]

Second, organizations have a culture of dominance when interpersonal relations are antagonistic and there is a lack of acceptance of cultural dissonance; characteristically, members hold the "belief that they are superior" to others. Typically homogeneous, traditional, and past oriented, dominance cultures also tend toward subordinating an organization's activities to proscribed cultural models and exhibit an aversion to uncertainty in favor of security.[82] Because of the fundamental nature of resistance and the groups that practice it, cultures of dominance should theoretically make up the vast majority of resistance organizations and movements, particularly those that use violent tactics.

Third, in a culture of adaptation, the organization remains unaccepting of cultural dissonance, but interpersonal relationships are nonantagonistic. "Oriented towards non-routine activities in a competitive environment," adaptation cultures place significant weight on the quality of interpersonal bonds and communication, the formal aims of the organization, and motivation for achievements. Tolerant of uncertainty and future oriented, adaptive cultures are "homogenous in the area of values" and exhibit partner-like power relationships.[83]

Finally, organizational cultures of cooperation (likely the least common organizational culture among resistance groups), with both the acceptance of cultural dissonance and nonantagonistic interpersonal relations, are characterized by "the rule of harmony with the environment" and the avoidance of "conflicts and rivalry." Instead, these heterogenic cultures employ democratic management styles that value personal bonds and dialogue under "the rules of autonomy, equality, [and] respect for differences."[84]

Environmental factors may significantly influence how the culture of a resistance organization develops, and many organizational facets may be dependent on, or causal toward, the organizational culture of the resistance group. In particular, security demands forced on both violent and nonviolent resistance movements to counter their opponents can deeply impact the culture of an organization. Such impacts result from endemic distrust and lead to dominance cultures. Cultural dissonance is often discouraged as threatening in resistance, as some differences may be perceived as signs of disloyalty. This potential dynamic is magnified in demanding security environments, as the fear of discovery or espionage can breed distrust and antagonism, affecting participants' interpersonal relationships.

Organization Strategic Dilemmas

The organization attribute presents at least six significant strategic dilemmas to resistance leaders. First, the organization dilemma refers to the pitfalls and benefits of the movement's formal bureaucratization. Although formal, hierarchic, and centralized structures may produce more coordinated efforts and legitimacy in the public sphere, they may also entail the abandonment of the movement's core goals or alienation of the grass roots.[85] Second, the band-of-brothers dilemma concerns the idea that although "affective loyalties to the broader group are essential," dependence on this dynamic for cohesion may become the lone incentive for continued participation "at the expense of the larger collectivity."[86] Third, the money's curse dilemma describes the tendency of some ideological social movements to view money as "dirtying [their] hands, yet even organizations that are 'above' such mundane issues nonetheless depend on financial resources,"[87] presenting the dilemma of how openly or shamelessly the group addresses its monetary needs.

The extension dilemma is the problem presented by the movement's growth: the further the group expands, "the less coherent [its] goals and actions can be." Additionally, although growth can increase a movement's power and reach, it also magnifies coordination problems. The potential for internal tensions also increases with group expansion.[88] Typologies related to the extension dilemma include membership strategies and the concept of replication. An adaptation of the extension dilemma is that of the ambitious leader, which stipulates that although a resistance may "want strong and competent leaders," there is a potential that "if they are too ambitious they may substitute their own goals for those of the group."[89] Finally, the dilemma of leadership distance (directly related to the theory of organization and leadership characteristic typologies) poses the question of whether the leadership will "be more appealing" as either a "lofty and unique . . . superhuman saint" or "a regular type, one of the guys."[90]

Security Environment

The security environment constitutes the procedural, communication, and structural measures imposed on participants and enforced by the resistance leadership in order to protect the activities of the group or movement from penetration by, or exposure to, its opponents. The overall security environment can include the compartmentalization of information between functional cells, the use of intermediary communication methods and recognition signals, the limiting of the size of functional cells, organizational redundancy through the maintenance of parallel cells, the screening of new members, and the inclusion of oaths and other checks on the loyalty of participants to prevent penetration by opponents.[91]

A few factors relevant to the security environment are considered elsewhere in this ARIS typology, and nearly all organizational issues have security implications. Specifically, oaths and loyalty checks are considered in the "Discipline and Rules" section describing a one-dimensional typology. Likewise, screening new members is effectively addressed in the variables in the columns in Table 12. The typology of security environments presented here forgoes these factors, instead examining the conceptual management of information and communication within the group (see Table 14).

Organized two-dimensionally, the typology of security environments is a product of the separation between functional cells within the organization and the depth of intermediary communications between cells and either leadership or other cells. Functional cells within the resistance organization may be isolated from one another, unaware of either the membership of other portions of the organization or of the existence of other functional cells. On the other hand, cells may be networked and aware of each other, able to communicate laterally rather than only through leadership. Their use of intermediary communication techniques (such as mail drops or anonymous messengers) can be categorized as either limited (a single point or not used at all) or extensive (layered and/or redundant use of intermediaries).

Table 14. Types of security environments.

	Separation of Functional Cells	
	Isolated	Networked
Use of Intermediary Communication		
Limited (none or single)	Secured cells	Unprotected network
Extensive (layered/redundant)	Highly secured cells	Protected network

First, an unprotected network exists when the resistance organization allows for lateral networking between functional cells and only enforces the limited use of intermediary communications. Second, a protected network exists when laterally connected functional cells ensure the extensive use of intermediary communication precautions. Third, a resistance organization has created secured cells (also known as "cells-in-series")[92] when the cells are isolated from one another but they make only limited use of intermediary communications. Finally, highly secured cells exist when they are isolated and secured through multiple layers of intermediary communication links. ("Parallel cells," redundant functional cells that are isolated in case one is compromised by opponents, would also constitute a security environment composed of highly secured cells.)[93]

ACTIONS

The actions attribute is concerned with the means by which actors carry out resistance as they engage in behaviors and activities in opposition to a resisted structure. Actions can encompass both the specific tactics used by a resistance movement and the broader characteristics or repertoires for action (i.e., strategy; see Figure 11).

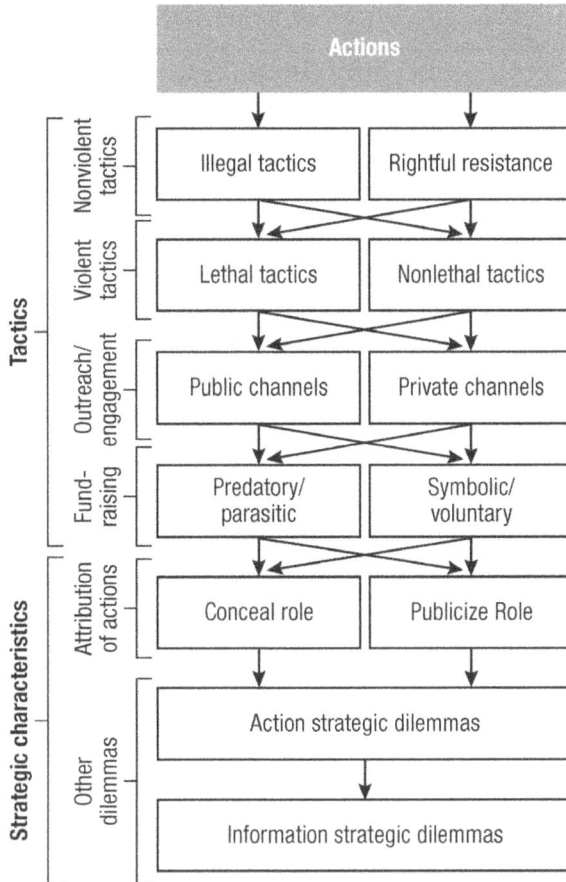

Figure 11. Actions kind hierarchy.

Tactics

Tactics are the methods and acts through which participants carry out resistance. As illustrated in the two-dimensional typology presented here (see Table 15), the types of tactics leveraged by a resistance group or movement can be legal or illegal, violent or nonviolent. When considering legality, it is important to distinguish between jurisdictions. There is only a very narrow aperture for legal nonviolent tactics (to the point that conceptualizing them as resistance tactics is a topic of debate among scholars), and violent tactics can be considered legal only when the conflict escalates past domestic jurisdiction into the international laws of war. *Violence* is used here to mean physical force to kill, hurt, or damage, as opposed to structural, direct political, symbolic, or everyday conceptualizations of violence.

Table 15. Types of resistance tactics.

	Legality (Do the Tactics Violate the Law?)	
	Legal	**Illegal**
Violence (Do the Tactics Inflict Physical Harm?)		
Violent	Military operations of a belligerency in accordance with international laws of war	Violent tactics
Nonviolent	Rightful resistance tactics	Nonviolent tactics

Nonviolent resistance tactics are often illegal, but those that fall within the legal parameters can be characterized as rightful resistance tactics (a distinction in resistance proposed by Kevin J. O'Brien in his 1996 *World Politics* article, "Rightful Resistance").[94] Many rightful resistance tactics are arguably identical to those used in conventional political and activist efforts. Nevertheless, they are included in this typology for two reasons. First, all resistance movements have these tactics at their disposal and likely use them in conjunction with other types of actions. Second, rightful resistance tactics uniquely operate "near the boundary of an authorized channel" and hinge "on locating and exploiting divisions among the powerful,"[95] thus maintaining the disruptive and subversive nature of resistance as a phenomenon.

The vast majority of violent resistance tactics are illegal, but some militarized political movements may be permitted to implement violent tactics that are considered legal in the context of the international laws of war (although they will continue to be illicit under domestic laws). This distinction applies when such movements progress to the point of recognition as an insurgency, where there exists a recognized non-international armed conflict (civil war), or are considered belligerents (in which case the conflict looks more like traditional interstate war and the belligerents control substantial territory). Violent tactics are permitted in these cases because in a noninternational armed conflict or an international armed conflict, the law of war applies (but different provisions are applicable depending on the nature of the conflict).

Violent Tactics

The use of physical violence by resistance movements can be further typified according to the lethality of the tactics. Lethal violent tactics are those intended or reasonably expected to result in the death of target or collateral persons. The most obvious of these in the history of resistance movements are the paramilitary operations of guerrilla and insurgent groups, including but not limited to small-arms and light-weapons combat and the use of explosive mines and improvised explosive devices, mortar and rocket attacks, and other lethal weapons systems in a framework similar to that of conventional military operations. Targeted killings and assassinations would also be considered a lethal tactic. Resistance movements willing to use lethal violent tactics, often at an asymmetric disadvantage in numbers and resources, may resort to terrorism. Such measures can be either targeted or indiscriminate.

Nonlethal violent tactics are those that inflict physical harm to persons or property with the reasonable expectation that the violent action will not result in loss of human life. Such tactics include violent demonstrations in the form of nonlethal riots and mobs (although provocations can escalate such actions into lethal confrontations), as well as various forms of nonlethal crime and intimidation (such as extortion and kidnapping). Sexual violence is another demonstrated tactic in this subset. Additionally, some violent means and methods often considered lethal could be used in a nonlethal manner. Examples include bombing without killing and the demonstration of lethal capabilities while avoiding loss of life.[96]

Nonviolent and Rightful Resistance Tactics

In his book *The Politics of Nonviolent Action*, scholar Gene Sharp presented a thorough accounting of the types of nonviolent resistance, which became widely known as his 198 methods.[97] Although they have been presented elsewhere in the ARIS body of work,[98] Sharp's 198 methods are reproduced here because they are typologically exhaustive of the nonviolent and rightful resistance tactics available to movements, groups, and participants. There has been no effort here to differentiate between particular methods as either legal (rightful resistance) or illegal (nonviolent tactic), as this categorization will differ for every national and local context according to the laws in place, and such detailed differentiation can take place on a case-by-case basis in the coding of data for research.

Although certainly exhaustive when originally written in 1973, the advent of the Internet and the widespread availability of innovative information technologies have dramatically changed the face of resistance movements and created a vast number of completely new tactics. For the thorough and effective analysis of modern cases of resistance, there is a glaring need to expand Sharp's methods to include both methods of cyber protest and noncooperation and methods of nonviolent cyber intervention (see Table 16).

Table 16. Types of nonviolent and rightful resistance tactics.

Methods of Nonviolent Protest and Persuasion	
Formal Statements	Public speeches
	Letters of opposition or support
	Declarations by organizations and institutions
	Signed public statements
	Declarations of indictment and intention
	Group or mass petitions
Communications with a Wider Audience	Slogans, caricatures, and symbols
	Banners, posters, and displayed communications
	Leaflets, pamphlets, and books
	Newspapers and journals
	Records, radio, and television
	Sky/earth writing
Group Representations	Deputations
	Mock awards
	Group lobbying
	Picketing
	Mock elections
Symbolic Public Acts	Displays of flags and symbolic colors
	Wearing of symbols
	Prayer and worship
	Delivering symbolic objects
	Protest disrobings
	Destruction of own property
	Symbolic lights

Methods of Nonviolent Protest and Persuasion (continued)	
Symbolic Public Acts (continued)	Displays of portraits
	Paint as protest
	New signs and names
	Symbolic sounds
	Symbolic reclamations
	Rude gestures
Pressures on Individuals	"Haunting" officials
	Taunting officials
	Fraternization
	Vigils
Drama and Music	Humorous skits and pranks
	Performances of plays and music
	Singing
Processions	Marches
	Parades
	Religious processions
	Pilgrimages
	Motorcades
Honoring the Dead	Political mourning
	Mock funerals
	Demonstrative funerals
	Homage at burial places
Public Assemblies	Assemblies of protest or support
	Protest meetings
	Camouflaged meetings of protest
	Teach-ins
Withdrawal and Renunciation	Walk-outs
	Silence
	Renouncing honors
	Turning one's back
Methods of Social Noncooperation	
Ostracism of Persons	Social boycott
	Selective social boycott
	Lysistratic nonaction

Methods of Social Noncooperation (continued)	
Ostracism of Persons (continued)	Excommunication
	Interdict
Noncooperation with Social Events, Customs, and Institutions	Suspension of social and sports activities
	Boycott of social affairs
	Student strike
	Social disobedience
	Withdrawal from social institutions
Withdrawal from the Social System	Stay-at-home
	Total personal noncooperation
	"Flight" of workers
	Sanctuary
	Collective disappearance
	Protest emigration (*hijrat*)
Methods of Economic Noncooperation: Economic Boycotts	
Actions by Consumers	Consumers' boycott
	Nonconsumption of boycotted goods
	Policy of austerity
	Rent withholding
	Refusal to rent
	National consumers' boycott
	International consumers' boycott
Action by Workers and Producers	Workman's boycott
	Producers' boycott
Action by Middlemen	Suppliers' and handlers' boycott
Action by Owners and Management	Traders' boycott
	Refusal to let or sell property
	Lockout
	Refusal of industrial assistance
	Merchants' "general strike"
Action by Holders of Financial Resources	Withdrawal of bank deposits
	Refusal to pay fees, dues, and assessments
	Refusal to pay debts or interest
	Severance of funds and credit
	Revenue refusal
	Refusal of a government's money

Methods of Economic Noncooperation: Economic Boycotts (continued)	
Actions by Governments	Domestic embargo
	Blacklisting of traders
	International sellers' embargo
	International buyers' embargo
	International trade embargo
Methods of Economic Noncooperation: The Strike	
Symbolic Strikes	Protest strike
	Quickie walkout (lightning strike)
Agricultural Strikes	Peasant strike
	Farm workers' strike
Strikes by Special Groups	Refusal of impressed labor
	Prisoners' strike
	Craft strike
	Professional strike
Ordinary Industrial Strikes	Establishment strike
	Industry strike
	Sympathetic strike
Restricted Strikes	Detailed strike
	Bumper strike
	Slowdown strike
	Working-to-rule strike
	Reporting "sick" (sick-in)
	Strike by resignation
	Limited strike
	Selective strike
Multi-Industry Strikes	Generalized strike
	General strike
Combination of Strikes and Economic Closures	Hartal
	Economic shutdown
Methods of Political Noncooperation	
Rejection of Authority	Withholding or withdrawal of allegiance
	Refusal of public support
	Literature and speeches advocating resistance

Methods of Political Noncooperation (continued)	
Citizens' Non-cooperation with Government	Boycott of legislative bodies
	Boycott of elections
	Boycott of government employment and positions
	Boycott of government departments, agencies, and other bodies
	Boycott of government-supported organizations
	Refusal of assistance to enforcement agents
	Removal of own signs and placemarks
	Refusal to accept appointed officials
	Refusal to dissolve existing institutions
Citizens' Alternatives to Obedience	Reluctant and slow compliance
	Nonobedience in absence of direct supervision
	Popular nonobedience
	Disguised nonobedience
	Refusal of an assemblage or meeting to disperse
	Sitdown
	Noncooperation with conscription and deportation
	Hiding, escape, and false identities
	Civil disobedience of "illegitimate" laws
Action by Government Personnel	Selective refusal of assistance by government aides
	Blocking of lines of command and information
	Stalling and obstruction
	General administrative noncooperation
	Judicial noncooperation
	Deliberate inefficiency and selective noncooperation by enforcement agents
	Mutiny
Domestic Governmental Action	Quasi-legal evasions and delays
	Noncooperation by constituent governmental units
International Governmental Action	Changes in diplomatic and other representations
	Delay and cancellation of diplomatic events
	Withholding of diplomatic recognition

Methods of Political Noncooperation (continued)	
International Governmental Action (continued)	Severance of diplomatic relations
	Withdrawal from international organizations
	Refusal of membership in international bodies
	Expulsion from international organizations
Methods of Nonviolent Intervention	
Psychological Intervention	Self-exposure to the elements
	The fast (fast of moral pressure, hunger strike, or satyagrahic fast)
	Reverse trial
	Nonviolent harassment
Physical Intervention	Sit-in
	Stand-in
	Ride-in
	Wade-in
	Mill-in
	Pray-in
	Nonviolent raids
	Nonviolent air raids
	Nonviolent invasion
	Nonviolent interjection
	Nonviolent obstruction
	Nonviolent occupation
Social Intervention	Establishing new social patterns
	Overloading of facilities
	Stall-in
	Speak-in
	Guerrilla theater
	Alternative social institutions
	Alternative communication system
Economic Intervention	Reverse strike
	Stay-in strike
	Nonviolent land seizure
	Defiance of blockades

Methods of Nonviolent Intervention (continued)	
Economic Intervention (continued)	Politically motivated counterfeiting
	Preclusive purchasing
	Seizure of assets
	Dumping
	Selective patronage
	Alternative markets
	Alternative transportation systems
	Alternative economic institutions
Political Intervention	Overloading of administrative systems
	Disclosing identities of secret agents
	Seeking imprisonment
	Civil disobedience of "neutral" laws
	Work-on without collaboration
	Dual sovereignty and parallel government

Fund-Raising

Resistance organizations and movements must raise funds for their operations if they are to achieve their objectives and functional longevity. R. T. Naylor analyzed the illicit fund-raising activities of specifically violent resistance groups in his *Crime, Law and Social Change* article, "The Insurgent Economy: Black Market Operations of Guerrilla Organizations," and presented three typological fund-raising categories: predatory, parasitical, and symbiotic.

Predatory resistance fund-raising, Naylor contests, consists of primarily "once-for-all [or one-off] activities," largely akin to "simple blue-collar criminal activities" such as robbery, ransom kidnapping, maritime fraud, and counterfeiting.[99] Second, parasitical fund-raising is characterized as "on-going sources like embezzlment [*sic*] and extortion . . . more akin to the activities of 'organized crime' syndicates."[100] While this type can take the shape of "revolutionary taxation" levied "on the income or wealth of well-to-do individuals or businesses," it can also emerge as more conventional "protection" fees imposed on smugglers, trade groups, or others.[101] Third, symbiotic fund-raising emerges when the resistance group gains the "capacity to profiteer . . . from the impact of its own [activities]," establishing its own profitable enterprises (both legal and illicit, including the production, smuggling, and sale of narcotics), as well as potential "parallel taxation" or "foreign trade taxes" to support its continued activities.[102]

Although typologically incisive and useful, Naylor's three categories of resistance fund-raising are nevertheless particular to violent insurgencies and thus require some slight addition and revision for use in the study of resistance as a whole. To this end, voluntary fund-raising, by which a group or movement acquires revenue through the solicitation of potentially sympathetic individuals, demographics, or organizations for financial support, may be considered a fourth category alongside the three types detailed above. Voluntary efforts would likely include both publicly and privately targeted messaging efforts to convince (rather than compel) financial support. Similarly, although predatory and parasitical fund-raising activities are almost exclusively reserved to those groups willing to use or threaten violence, symbiotic efforts can conceivably exist in the realm of nonviolent resistance via the provision of services, the quid pro quo offer of messaging or political favors in return for voluntary contributions, and many other symbiotic relationships.

Table 17. Types of resistance fund-raising.

	Original Source of Wealth Creation		
	Labor of Those outside the Resistance Group	Labor of the Resistance Group	Labor of Resistance Supporters
Nature of Funding Actions			
Once-for-all (one-off)	Predatory	Symbiotic	Voluntary
Ongoing	Parasitic		

These four types of resistance fund-raising (predatory, parasitic, symbiotic, and voluntary) can be conceptually deconstructed and categorized in a two-dimensional typology (see Table 17). The defining characteristics are the original source of wealth creation (either those outside or within the movement) and the nature of funding actions (one-off or ongoing). Fund-raising efforts that draw from the wealth of those outside the movement are differentiated between predatory and parasitic according to their one-off or ongoing nature. However, symbiotic and voluntary fund-raising can be of either nature. Instead, they are specified according to the source of wealth: symbiotic fund-raising is direct wealth creation by the resistance group, while voluntary fund-raising draws from the contributions of those supporters outside the group itself.

Strategic Characteristics

The strategic characteristics of actions are the aspects over which the resistance group or movement can either practice agency or attempt to do so. This agency, or strategic choice, is practiced in regard to attribution, outreach, and other strategic dilemmas concerning actions and information operations. However, some discrete types (see attribution of actions) apply when the agency of the resistance is overcome by external factors.

Attribution of Actions

The attribution of resistance actions is a critical strategic characteristic, particularly for violent movements that might seek to either take credit for successful strikes against opponents or, conversely, avoid culpability for potentially unpopular operations or gaffes. An original conceptual typology of attribution can be characterized in two dimensions, one being the resistance organization's strategic intent regarding attribution and the other being the public's or opponent's knowledge of the resistance movement's role in the given action (see Table 18), resulting in four types: clandestine operations, exposed operations, public operations, and suppressed or ignored operations.

First, clandestine operations are those in which the role of the resistance organization is successfully concealed from the public and opponent. Suspicions about the resistance group's involvement may exist, but the group does not openly take credit for the action and may actively deny involvement. Second, exposed operations are those in which the resistance tried to conceal its role in a given action but was exposed by its opponents or other external actors. This type can take shape as blame for the consequences of an action, the prosecution of participants for their roles, or the propagation of a conspicuous failure.

Third, public operations are those in which the resistance group readily and publicly accepts attribution for given actions. This type will often take shape as either open identification as participants in a given group during the given action or announcements claiming attribution for the action. Finally, suppressed or ignored operations are those in which the resistance group seeks to propagate its role in a given action but that attribution is overshadowed, with knowledge of the group's role either suppressed or diverted to another actor.

Table 18. Types of action attribution.

	Strategic Intent of Resistance Organization	
	Conceal Role in Action	Publicize Role in Action
Public and/or Opponent Knowledge of Resistance Organization Role in Action		
Aware of resistance role in action	Exposed operations	Public operations
Unaware of resistance role in action	Clandestine operations	Suppressed or ignored operations

Outreach and Engagement

Resistance groups and organizations must be able to communicate with the public and potential participants in order to succeed or persevere, necessitating the use of various means of outreach and engagement. David A. Snow, Louis A. Zurcher Jr., and Sheldon Ekland-Olson contest that outreach and engagement serve three purposes for social movements (information dissemination and operations, movement and organization promotion, and movement or organization recruitment) and propose a "classification of general outreach and engagement possibilities for movement information dissemination, promotion, and recruitment." This classification is adapted below as a two-dimensional typology (see Table 19).[103]

Table 19. Types of outreach and engagement.

	Visibility of Communication Channels	
	Public	Private
Intimacy of Engagement		
Face to face	Direct public	Direct network
Mediated	Intermediate public	Intermediate network

The two defining factors of this typology of outreach and engagement methods are the intimacy of the engagement ("whether they are face to face or mediated") and the openness or visibility of that engagement ("a continuum ranging from public to private"). More specifically, face to face is defined as "all information, whether it be verbal or nonverbal, that is imparted when two or more individuals or groups are physically present," whereas mediated engagement "refers to information dissemination through institutionalized mass communication mechanisms . . . or through institutionalized, but individualized and privatized, communication mechanisms."[104] The intersection of these factors results in four typological categories of outreach and engagement methods: direct public, intermediate public, direct network, and intermediate network.

First, direct public engagement involves face-to-face interaction in an open setting, including public "leafleting, petitioning, and proselytizing . . . participation in public events . . . [and staged] events for public consumption, such as sit-ins, protests . . . conventions and festivals." Second, intermediate public engagement primarily entails the use of mass media for broad dissemination (Internet, television, radio, and newspaper). Third, direct network engagement involves "information dissemination and recruitment" through either "door-to-door leafleting, petitioning, and proselytizing" or "among familiar others along the lines of . . . extra-movement interpersonal networks." Finally, intermediate network engagement is conducted through ostensibly private communications, including mail, telephone, and private Internet communications.[105]

Action and Information Strategic Dilemmas

Most strategic dilemmas may be categorized under the actions attribute and subdivided into action and information dilemmas, where the former concern the decision on and coordination of certain tactics and the latter address propaganda, public relations, and other issues related to public appeals.

Action Dilemmas

Six strategic dilemmas particularly concern the actions and tactics of a resistance group or movement. First, the direct or indirect moves dilemma concerns the strategic weight of confrontations, which can either be "direct confrontations with opponents" or "indirect moves such as persuading third parties, gathering resources, building networks, and so on."[106] A group may choose consistent and open confrontations, a more clandestine or subtle rallying of support and resources, or a mix of the two. Second, the plans versus opportunity dilemma determines whether a group or movement plans initiatives of its own or "wait[s] for opponents to make mistakes" before finally taking action.[107] Planned initiatives allow for strategic latitude of choice and strong preparation but also carry the risk of conspicuous failure that may undermine the viability of the movement. A lie-in-wait strategy, however, allows for a higher likelihood of success because the group acts in the opponent's moment of weakness, but this strategy carries the risk of allowing the group or movement to fall into crippling obscurity.

Third, the basket dilemma may be characterized as the choice between "one decisive engagement" in a "winner-take-all" ultimatum and spreading risk "over many small engagements" with the opponent of resistance.[108] Although the majority of resistance movements are small challengers that take the path of dispersed risk through asymmetric conflict, some may choose decisive engagements. Fourth, the engagement dilemma concerns the decision to move from a state of latency to open resistance activity, given that "visibility brings a number of risks, such as external repression or misrepresentation and internal conflicts over strategy."[109]

Fifth, the dirty hands dilemma largely concerns a resistance movement's transition from using primarily rightful resistance tactics to leveraging illicit and/or violent tactics as well, recognizing that "some goals are only—or more easily—attained through unsavory means."[110] Finally, the familiar and the new dilemma concerns the fact that although "new tactics surprise opponents and authorities . . . it is typically hard for [a resistance] group to pull them off."[111] Tactical innovation may provide a strategic edge, but it may also constitute a stumbling block for the resistance group by complicating planning and logistics.

Information Dilemmas

Because perceptions and the rhetorical contest of propaganda form much of any resistance group's strategic center of gravity, the information and messaging sphere provides the greatest variety of strategic dilemmas for consideration by a resistance effort (totaling nine). First, the form or content dilemma is the option between using "procedural rhetoric" and "substantive rhetoric." While the resistance can "switch attention and rhetoric from the content of [claims] to the formal mechanisms for handling [them]," such information maneuvers heighten the risk of "losing sight of the original issue."[112] Second, the reaching-out or reaching-in dilemma concerns how actions are oriented to participants in resistance or to outsiders. Related to the extension dilemma, reaching in may serve to provide symbolic or material rewards to preserve a sense of exclusivity, homogeneity, and/or doctrinal purity as the movement seeks to expand.[113]

Third, the dilemma of cultural innovation addresses the tension between the change the resistance seeks and the group's need to appeal to various audiences by using "the meanings they already hold, [where] pushing too far may cause" alienation and undermine messaging campaigns.[114] Fourth, the victim or hero dilemma concerns how the resistance group portrays itself in messaging, either "as wronged victim in need of help or as [a] strong, avenging hero."[115] Each role may be useful at different times, depending on the need to inspire confidence or attract external support. Similarly, the villain or clown dilemma concerns the strategic portrayal of resistance opponents, either "as strong and dangerous or as silly and contemptible."[116]

Sixth, the radical-flank dilemma concerns whether or not to radicalize rhetoric, tactics, or both, given that "extreme words and actions get attention, and often take opponents by surprise, but they usually play poorly with bystanders and authorities."[117] Seventh, the media dilemma addresses the choice of whether to depend on popular media as an avenue to propagate the resistance message, considering that popular media sources reach "broad audiences, but . . . they are likely to distort [the message] in doing so."[118] Eighth, the bridge-builders dilemma highlights the choice whether to leverage "individuals who can mediate between groups, or different sides in a conflict," given that such actors or leaders "often lose the trust of their own groups by doing so."[119] Finally, the segregating audiences dilemma concerns the strategic desire of the resistance "to send different messages to different players." Such dual messaging makes the group vulnerable to counter-messaging where both messages "can be used to make [the group] look duplicitous."[120]

CONCLUSION

Resistance is a phenomenon of human behavior, society, and politics that is broader than formal insurgencies and self-proclaimed revolutions. Nonviolent contentious movements have had increasingly high-profile effects on regimes and world affairs, as demonstrated by the Arab Spring and Euromaidan. The conceptual desegregation of resistance for holistic study continues to progress, and rightly so. USASOC and the ARIS team hope that this typology, in conjunction with other efforts, will make worthwhile contributions to future research into this phenomenon. The ARIS typology can be an instrument that is adapted, built on, and deviated from as required for the shared conceptualization of resistance in all its manifestations, thus providing a resource for robust interdisciplinary study into the phenomenon as a whole, rather than only its parts.

GLOSSARY

Actions: The means by which actors carry out resistance as they engage in behaviors and activities in opposition to a resisted structure. Actions can encompass both the specific tactics used by a resistance movement and the broader characteristics or repertoires for action (i.e., strategy).

Actors: The individual and potential participants in an organized resistance, as well as external contributors and either competing or cooperating resistance groups.

Actors, Direct: Those actors who will inevitably emerge within any resistance group or movement, namely the leadership and participants thereof.

Actors, Indirect: Those actors who indirectly impact or may emerge in or around a resistance movement, including the general population, other resistance groups, and external supporters.

Administrator-Executive: A leader who implements the policies of the movement, completing the formal institutionalization of the goals of the resistance.[121] (See Leadership.)

Agitator: A leader who "stirs the people not by what he does, but by what he says . . . [leading] people to challenge and question . . . [the status quo and] create unrest," or who serves "to intensify, release, and direct tensions which people already have."[122] (See Leadership.)

Alterative Rationale: A resistance rationale that seeks limited or specified change among some individuals or a particular community, often concerning the way people think about certain behaviors or issues within a given system or structure. (See Rationale.)

Attribute: An inherent or fundamental characteristic.

Auxiliary: Also referred to as supporters, the auxiliary are nonmartial participants who actively support the resistance effort on a part-time basis and on the periphery of their work and lifestyle.

Causes: The collectively expressed rationales for resistance and the individual motivations for participation.

Characteristic: A distinguishing trait or quality.

Classic Counterinsurgency: A counterinsurgency model "most often associated with . . . highly capable Western democracies" that seeks to accommodate and provide public goods while limiting or carefully targeting any use of violence.[123] (See Counterinsurgency.)

Concept: "An idea of a phenomenon formed by combining its attributes . . . alternatively, a mental image that, when operationalized, helps to organize the data analysis."[124]

Concept of Replication: Facilitates a strategy for a movement's growth outside the core organization through the foundation of affiliated parties or groups or the adoption of existing groups into the movement.

Containment: A counterinsurgency model in which a state is "unable or unwilling to accommodate the reconcilable opposition" and "uses force—usually indiscriminately—to repress insurgent activity," accepting risk of residual violence as long as the resistance is relegated to certain, often poorly defined, bounds.[125] (See Counterinsurgency.)

Core Membership: See Underground.

Counterinsurgency: The operations and policies of a government entity "designed to simultaneously defeat and contain insurgency" and potentially "address its root causes."[126]

Environment: The preexisting and emerging conditions within the political, social, physical, or interpersonal contexts that enable or constrain the mobilization of resistance, directly or indirectly.

External Support: The provision of support to a resistance group or movement by entities or individuals external to the country.

Fighters: Those participants "organized along military lines to conduct military and paramilitary operations" in a resistance but also including those similarly organized for other forms of "subversion and violence."[127]

Founding Members: Participants who contribute to the resistance group or movement in its earliest stages.

Informal Accommodation: A counterinsurgency model in which the state is unable or unwilling to provide public goods to discontented communities but is willing to compromise with components of the resistance.[128] (See Counterinsurgency.)

Kind Hierarchy: "An ordered relationship among concepts, in which subordinate concepts may be understood as 'a kind of' in relation to superordinate concepts. This is a basic feature of conceptual structure, both in social science and in ordinary usage."[129]

Leadership: Those individuals within a movement or organization who "provide strategic and tactical direction, organization, and the ideology of the movement," performing "these functions within the unique and compelling context of their country, culture, and political economy."[130]

Moral Support: A nonmaterial form of passive or tacit external support, only going so far as sympathetic public statements or similar measures. (See External Support.)

Motivations: The sources of willingness for individuals to participate in a resistance, manifesting "as a function of the perceived attractiveness or aversiveness of the expected consequences [costs and benefits] of participation."[131]

Motives, Collective: A function of the personal "expectation that participation will help to produce [a] collective good and the value of [said] collective good."[132] (See Motivations.)

Motives, Reward: Willingness to participate in resistance derived from associated "non-social costs and benefits."[133] (See Motivations.)

Motives, Social: A function of the costs and benefits of participation "as distinguished in the reactions of significant others."[134] (See Motivations.)

Movement: The overall collective effort of resistance, potentially encompassing multiple organizations and other actors.

Organization: "The internal characteristics of a movement: its membership, policies, structures, and culture."[135]

Participants: The individuals acting in concert with, or in support of, a resistance movement or organization and its efforts.

Phenomenon: An event or series of events that can be observed and studied; used in reference to resistance.

Political Support: Active and nonmaterial external support to resistance, including advocacy and symbolic actions to express committed support. (See External Support.)

Prophet: A leader who "feels set apart or called to leadership" and claims "special and separate knowledge of the causes of unrest and discontent," speaking "with an air of authority . . . in general terms."[136] (See Leadership.)

Rationale: The collectively or organizationally propagated narrative for collective action, outlining the resistance movement's values, claims, and objectives.

Recruits: Participants in resistance who were incorporated into the group or movement once the effort was already under way.

Redemptive Rationale: A resistance rationale that seeks dramatic change among some individuals or a specified community, often in the form of a complete transformation of the specified person(s) or their circumstances. (See Rationale.)

Reformative Rationale: A resistance rationale that seeks a specific change or changes to the existing structure or system on a large scale, often characterizing the targeted change as key to improving society as a whole. (See Rationale.)

Reformer: A leader who "attacks specific evils and develops a clearly defined program," attempting "to change conditions in conformity with his own conceptions of what is good and desirable."[137] (See Leadership.)

Resistance: A form of contention or asymmetric conflict involving participants' limited or collective mobilization of subversive and/or disruptive efforts against an authority or structure.

Resources: External support for resistance through the active provision of material aid. (See External Support.)

Revolutionary Rationale: A resistance rationale that seeks to overthrow the standing system to which participants are subject. Depending on the movement's objectives, it may seek to establish a new structure in place of the old. This is often characterized as a transformation of society, wholly or partially. (See Rationale.)

Sanctuary: An external supporter's provision of passive material aid in the form of safe havens, secure training sites, operational bases, protection from extradition, or other shields from the adversary's actions. (See External Support.)

Sleeper Cell: Includes participants in resistance who are organized in a martial fashion but contribute on only a part-time basis peripheral to their normal work and lifestyle, likely dormant within a given population until called on to conduct resistance activities.

Statesman: A leader "able to formulate policies and [who] will attempt to carry social policy into practice" and "will propose the program which promises to resolve the issues and realize the objectives of which the people have become aware."[138] (See Leadership.)

Strategic Dilemmas: Explicit choices concerning implicit trade-offs resistance organizers and participants face regarding their rationale, organization, actions, and information operations.[139]

Strong-State Repression: A counterinsurgency model in which the state emphasizes the provision of public goods to assuage discontent but is nevertheless "unable or unwilling to offer accommodations to any part of the opposition, relying instead on broad, often indiscriminate use of force and terror to suppress organized dissent."[140] (See Counterinsurgency.)

Structural Focus: The portion or portions of society in which the resistance movement or group seeks to enact change, as characterized in the rationale.

Supporter: See Auxiliary.

Tactics: The methods and actions through which participants carry out resistance.

Tactics, Nonviolent: Illicit actions and methods in resistance that do not inflict physical harm on people or property. (See Tactics.)

Tactics, Rightful Resistance: Actions and methods in resistance that uniquely operate "near the boundary of an authorized channel," remaining legal while hinging "on locating and exploiting divisions among the powerful,"[141] thus maintaining the disruptive and subversive nature of resistance as a phenomenon. (See Tactics.)

Tactics, Violent: Actions and methods in resistance that inflict physical harm on people or property. (See Tactics.)

Typology: "An organized system of types that breaks down an overarching concept into component dimensions and types."[142]

Typology, Conceptual: "A form of typology that explicates the meaning of a concept by mapping out its dimensions, which correspond to the rows and columns in the typology. The cell types are defined by their position vis-à-vis the rows and columns. May also be called a descriptive typology."[143]

Underground: Also referred to as core membership, the underground describes those participants integrated into at least one facet of resistance organization operations on a full-time basis.

NOTES

[1] David Collier, Jody LaPorte, and Jason Seawright, "Typologies: Forming Concepts and Creating Categorical Variables," in *The Oxford Handbook of Political Methodology*, ed. Janet M. Box-Steffensmeier, Henry E. Brady, and David Collier (Oxford: Oxford University Press, 2008).

[2] David Collier, Jody LaPorte, and Jason Seawright, "Putting Typologies to Work: Concept Formation, Measurement, and Analytic Rigor," *Political Research Quarterly* 65, no. 1 (2012): Appendix 2.

[3] Erin Hahn et al., *Developing a Typology of Resistance: A Structure to Understand the Phenomenon* (Fort Bragg, NC: United States Army Special Operations Command, 2015).

[4] Collier, LaPorte, and Seawright, "Putting Typologies to Work," 228.

[5] Collier, LaPorte, and Seawright, "Typologies."

[6] Ibid.; and Collier, LaPorte, and Seawright, "Putting Typologies to Work," 217–232.

[7] Collier, LaPorte, and Seawright, "Putting Typologies to Work," 223.

[8] Ibid., Appendix 1.

[9] Jeremy M. Weinstein, *Inside Rebellion: The Politics of Insurgent Violence* (Cambridge: Cambridge University Press, 2006), 19.

[10] Hahn et al., *Developing a Typology of Resistance*, 17–21.

[11] Robert Leonhard (ed.), *Undergrounds in Insurgent, Revolutionary, and Resistance Warfare*, 2nd ed. (Fort Bragg, NC: United States Army Special Operations Command, 2013), ix.

[12] Rex D. Hopper, "The Revolutionary Process: A Frame of Reference for the Study of Revolutionary Movements," *Social Forces* 28, no. 3 (March 1950): 272.

[13] Ibid., 275.

[14] Ibid.

[15] Ibid., 277.

[16] Ibid., 278–279.

[17] Ernesto Che Guevara, "The Cadres: Backbone of the Revolution," *Cuba Socialista*, September 1962, published online in 2002 to the Ernesto Che Guevara Internet Archive, https://www.marxists.org/archive/guevara/1962/09/misc/x01.htm.

[18] Cheryl Pellerin, "Lynn: Cyberspace is the New Domain of Warfare," *DoD News*, October 18, 2010, http://archive.defense.gov/news/newsarticle.aspx?id=61310.

[19] Andrew R. Molnar et al., *Undergrounds in Insurgent, Revolutionary, and Resistance Warfare* (Washington, DC: Special Operations Research Office, 1963), 371.

[20] Ibid., 47–126.

[21] Ibid., 365.

[22] Ibid., 47–126.

[23] US Joint Chiefs of Staff, *Department of Defense Dictionary of Military and Associated Terms*, Joint Publication 1-02 (JP 1-02) (Washington, DC: US Joint Chiefs of Staff, January 15, 2015), 103.

[24] Ibid., 119.

[25] US Department of the Army, *Tactics in Counterinsurgency*, Field Manual 3-24.2 (FM 3-24.2) (Washington, DC: Headquarters of the Department of the Army, April 2, 2009), 2–12.

[26] Ibid., 4–11.

27 *Merriam-Webster's Collegiate Dictionary*, 11th ed., s.v. "cause," http://www.merriam-webster.com/dictionary/cause.

28 Sydney G. Tarrow, *Power in Movement: Social Movements and Contentious Politics*, rev. and updated 3rd ed. (Cambridge: Cambridge University Press, 2011), 140–156; Patrick M. Regan, *Civil Wars and Foreign Powers: Outside Intervention in Intrastate Conflict* (Ann Arbor, MI: University of Michigan Press, 2000), 21–24; and Bruce Hoffman, *Inside Terrorism* (New York: Columbia University Press, 1998), 43.

29 Daniel Byman, "Outside Support for Insurgent Movements." *Studies in Conflict & Terrorism* 36, no. 12 (2013): 982.

30 Robert D. Benford and David A. Snow, "Framing Processes and Social Movements: An Overview and Assessment," *Annual Review of Sociology* 26 (2000): 611–639.

31 David F. Aberle, *The Peyote Religion among the Navaho*, 2nd ed. (Chicago: University of Chicago Press, 1982), 316.

32 Ibid.

33 Ibid., 316–317.

34 Bard E. O'Neill, *Insurgency & Terrorism: From Revolution to Apocalypse*, 2nd ed., rev. (Sterling, VA: Potomac Books, 2005); Christopher Alexander, Charles Kyle, and William S. McCallister, *The Iraqi Insurgent Movement* (Cambridge, MA: Commonwealth Institute, November 14, 2003); and Shelly Shah, "Social Movements: Meaning, Causes, Types, Revolution and Role," *Sociology Discussion* (online forum), 2013, http://www.sociologydiscussion.com/social-movements/social-movements-meaning-causes-types-revolution-and-role/2248.

35 O'Neill, *Insurgency & Terrorism*, 20–24; and Alexander et al., *Iraqi Insurgent Movement*, 3.

36 O'Neill, *Insurgency & Terrorism*, 26–28; and Alexander et al., *Iraqi Insurgent Movement*, 3.

37 Alexander et al., *Iraqi Insurgent Movement*, 3.

38 O'Neill, *Insurgency & Terrorism*, 24–26.

39 Shah, "Social Movements."

40 O'Neill, *Insurgency & Terrorism*, 28–29.

41 James M. Jasper, "A Strategic Approach to Collective Action: Looking for Agency in Social-Movement Choices," *Mobilization: An International Journal* 9, no. 1 (February 2004): 1.

42 Ibid., 8–9.

43 Ibid., 13.

44 Bert Klandermans, "Mobilization and Participation: Social-Psychological Expansions of Resource Mobilization Theory," *American Sociological Review* 49, no. 5 (October 1984): 586.

45 Ibid., 583–600.

46 Nathan Bos (ed.), *Human Factors Considerations of Undergrounds in Insurgencies*, 2nd ed. (Fort Bragg, NC: United States Army Special Operations Command, 2013), 132.

47 Klandermans, "Mobilization and Participation," 586.

48 Bos, *Human Factors*, 131–132, 134.

49 Klandermans, "Mobilization and Participation," 586.

50 Bos, *Human Factors*, 132–135.

51 Klandermans, "Mobilization and Participation," 586.

52 Bos, *Human Factors*, 131–132.

53 Ibid., 132–135.

[54] Christoph Reuter, "The Terror Strategist: Secret Files Reveal the Structure of Islamic State," *Spiegel Online*, April 18, 2015, http://www.spiegel.de/international/world/islamic-state-files-show-structure-of-islamist-terror-group-a-1029274.html.

[55] US Department of the Army, *Insurgencies and Countering Insurgencies*, Field Manual 3-24 (FM 3-24) (Washington, DC: Headquarters of the Department of the Army, May 2014), Glossary-2.

[56] Stephen Watts et al., *Countering Others' Insurgencies: Understanding U.S. Small-Footprint Interventions in Local Context* (Santa Monica, CA: RAND Corporation, 2014), 18.

[57] Ibid., 19.

[58] Ibid.

[59] Ibid., 19–20.

[60] Ibid., 20.

[61] Paul Staniland, "States, Insurgents, and Wartime Political Orders," *Perspectives on Politics* 10, no. 2 (June 2012): 243–264.

[62] Ibid., 247.

[63] Ibid., 248.

[64] Ibid., 249.

[65] Ibid., 250.

[66] Ibid., 251.

[67] Ibid., 252.

[68] Weinstein, *Inside Rebellion*, 19.

[69] Jurgen Willems and Marc Jegers, "Social Movement Structures in Relation to Goals and Forms of Action: An Exploratory Model," *Canadian Journal of Nonprofit & Social Economy Research (ANSESRJ)* 3, no. 2 (Autumn 2012): 67–81.

[70] Ibid., 71.

[71] Ibid.

[72] Ibid., 70.

[73] Ibid.

[74] Luther P. Gerlach, "The Structure of Social Movements: Environmental Activism and Its Opponents," in *Networks and Netwars: The Future of Terror, Crime, and Militancy*, ed. John Arquilla and David Ronfeldt (Santa Monica, CA: RAND Corporation, 2001), 289–310.

[75] Ibid., 289–290.

[76] Adapted from typologies proposed by Boaz Ganor, "Terrorist Organization Typologies and the Probability of a Boomerang Effect," *Studies in Conflict and Terrorism* 31, no. 4 (2008): 269–283.

[77] Leonhard, *Undergrounds*, 10–12; and Molnar et al., *Undergrounds*, 49–51.

[78] Olivier Bangerter, *Regulating Armed Groups from Within: A Typology*, Research Note 13 (Geneva: Small Arms Survey, January 2012).

[79] Quoted in Łukasz Sułkowski, "Typologies of Organisational Culture – Multi-dimentional Classifications," *Przedsiębiorczość i Zarządzanie* 14, no. 2 (2013): 180.

[80] Czesław Sikorski, *Kultura Organizacyjna* (Warsaw: C. H. Beck, 2002), quoted in Sułkowski, "Typologies," 173–182.

[81] Sułkowski, "Typologies," 181.

[82] Ibid.

83 Ibid.

84 Ibid.

85 Jasper, "Strategic Approach to Collective Action," 7.

86 Ibid., 13.

87 Ibid.

88 Ibid., 7–8.

89 Ibid., 13.

90 Ibid.

91 Leonhard, *Undergrounds*, 117–129; and Bos, *Human Factors*, 40–42.

92 Bos, *Human Factors*, 40–42.

93 Ibid.

94 Kevin J. O'Brien, "Rightful Resistance," *World Politics* 49, no. 1 (October 1996): 31–55.

95 Ibid., 33.

96 Roger D. Peterson, *Western Intervention in the Balkans: The Strategic Use of Emotion in Conflict* (Cambridge: Cambridge University Press, 2011), 17.

97 Gene Sharp, *The Politics of Nonviolent Action* (Boston: Portor Sargent, 1973), reproduced by the Albert Einstein Institution as "198 Methods of Nonviolent Action," http://www.aeinstein.org/wp-content/uploads/2013/09/198_methods-1.pdf.

98 Bos, *Human Factors*, 280–283.

99 R. T. Naylor, "The Insurgent Economy: Black Market Operations of Guerrilla Organizations," *Crime, Law and Social Change* 20, no. 1 (1993): 24–25, 28.

100 Ibid., 29.

101 Ibid., 29–30.

102 Ibid., 31–33.

103 David A. Snow, Louis A. Zurcher Jr., and Sheldon Ekland-Olson, "Social Networks and Social Movements: A Microstructural Approach to Differential Recruitment," *American Sociological Review* 45, no. 5 (October 1980): 789–790.

104 Ibid.

105 Ibid., 790.

106 Jasper, "Strategic Approach to Collective Action," 13.

107 Ibid.

108 Ibid.

109 Ibid.

110 Ibid.

111 Ibid.

112 Ibid., 9–10.

113 Ibid., 10.

114 Ibid., 13.

115 Ibid.

116 Ibid.

117 Ibid.

118 Ibid.

119 Ibid.

120 Ibid.

121 Hopper, "Revolutionary Process," 278–279.

122 Ibid., 272.

123 Watts et al., *Countering Others' Insurgencies*, 19.

124 Collier et al., "Putting Typologies to Work," Appendix 1.

125 Watts et al., *Countering Others' Insurgencies*, 20.

126 US Department of the Army, FM 3-24, Glossary-2.

127 US Joint Chiefs of Staff, JP 1-02, 103, 119.

128 Watts et al., *Countering Others' Insurgencies*, 19–20.

129 Collier et al., "Putting Typologies to Work," Appendix 1.

130 Leonhard, *Undergrounds*, ix.

131 Klandermans, "Mobilization and Participation," 586.

132 Ibid.

133 Ibid.

134 Ibid.

135 Weinstein, *Inside Rebellion*, 19.

136 Hopper, "Revolutionary Process," 275.

137 Ibid.

138 Ibid., 277.

139 Jasper, "Strategic Approach to Collective Action," 1.

140 Ibid., 19.

141 O'Brien, "Rightful Resistance," 33.

142 Collier et al., "Putting Typologies to Work," Appendix 2.

143 Ibid.

BIBLIOGRAPHY

Abdel Salam, El-fatih A. "A Typology of Conflicts and Conflict Resolution in the Muslim World." *Peace Research* 38, no. 1 (2006): 9–28.

Aberle, David F. *The Peyote Religion among the Navaho.* 2nd ed. Chicago: University of Chicago Press, 1982.

Abrahms, Max. "Why Terrorism Does Not Work." *International Security* 31, no. 2 (2006): 42–78.

Aguilar, Anthony B. "On Insurgency." Master's thesis, United States Army Command and General Staff College, 2008.

Alexander, Christopher, Charles Kyle, and William S. McCallister. *The Iraqi Insurgent Movement.* Cambridge, MA: Commonwealth Institute, November 14, 2003. http://www.comw.org/warreport/fulltext/03alexander.pdf.

Arendt, Hannah. *On Revolution.* New York: The Viking Press, 1963.

Armitage, David. "What's the Big Idea?" *The Times Literary Supplement*, September 20, 2012. http://www.the-tls.co.uk/tls/public/article1129685.ece.

Bangerter, Olivier. *Regulating Armed Groups from Within: A Typology.* Research Note 13. Geneva: Small Arms Survey, January 2012.

Barker, Chris. *Cultural Studies: Theory and Practice.* London: Sage, 2011.

Bass, Bernard M., and Bruce J. Avolio. "Transformational Leadership and Organizational Culture." *Public Administration Quarterly* 17, no. 1 (Spring 1993): 112–121.

Beckett, Ian F. W. *Modern Insurgencies and Counter-Insurgencies: Guerrillas and Their Opponents since 1750.* New York: Routledge, 2004.

Benford, Robert D., and David A. Snow. "Framing Processes and Social Movements: An Overview and Assessment." *Annual Review of Sociology* 26 (2000): 611–639.

———. "Ideology, Frame Resonance, and Participant Mobilization." *International Social Movement Research* 1, no. 1 (1998): 197–217.

Bennett, Richard R. "Presidential Address: Comparative Criminology and Criminal Justice Research: The State of Our Knowledge." *Justice Quarterly* 21, no. 1 (2004): 1–21.

Berlet, Chip, and Stanislav Vysotsky. "Overview of U.S. White Supremacist Groups." *Journal of Political and Military Sociology* 31, no. 1 (2006): 11–48.

Berman, Eli. *Radical, Religious, and Violent the New Economics of Terrorism.* Cambridge, MA: MIT Press (2009).

Blaser, Mario. "Ontological Conflicts and the Stories of Peoples in Spite of Europe: Toward a Conversation on Political Ontology." *Current Anthropology* 54, no. 5 (2013): 547–568.

Blattman, Christopher, and Edward Miguel. "Civil War." *Journal of Economic Literature* 48, no. 1 (2010): 3–57.

Bos, Nathan, ed. *Human Factors Considerations of Undergrounds in Insurgencies.* 2nd ed. Fort Bragg, NC: United States Army Special Operations Command, 2013.

Bourdieu, Pierre. *Language and Symbolic Power.* Cambridge, MA: Harvard University Press, 1993.

Bruce, Gary. *Resistance with the People: Repression and Resistance in Eastern Germany, 1945–1955.* Lanham, MD: Rowman and Littlefield Publishers, 2003.

Burgin, Martin. *Insurgency and Counterinsurgency as Historiographical Subjects: A Selected Bibliography.* Neuchâtel, Switzerland: International Commission of Military History, 2010. http://www.icmh.info/assets/userFiles/Ablage/Texte/Buergin,%20Insurgency%20and%20COIN.pdf.

Byman, Daniel. "Outside Support for Insurgent Movements." *Studies in Conflict & Terrorism* 36, no. 12 (2013): 981–1004.

———. *Understanding Proto-Insurgencies.* RAND Counterinsurgency Study Paper 3. Santa Monica, CA: RAND Corporation, 2007.

Campomanes, Oscar V. "Images of Filipino Racialization in the Anthropological Laboratories of the American Empire: The Case of Daniel Folkmar." *PMLA* 123, no. 5 (2008): 1692–1699.

Caygill, Howard. *On Resistance: A Philosophy of Defiance.* London: Bloomsbury Academic, 2015.

Center for the Study of Foreign Affairs. *Low-Intensity Conflict: Support for Democratic Resistance Movements—Report on a Colloquium Co-Sponsored by the Center for the Study of Foreign Affairs, Foreign Service Institute, and the United States Army Command and General Staff College, 7 January 1988.* Washington, DC: Foreign Service Institute; US Department of State, 1988.

Chasdi, Richard J. "Middle East Terrorism 1968–1993: An Empirical Analysis of Terrorist Group-Type Behavior." *The Journal of Conflict Studies* 17, no. 2 (1997): 1–4.

Chenoweth, Erica, and Orion A. Lewis. "Unpacking Nonviolent Campaigns: Introducing the NAVCO 2.0 Dataset." *Journal of Peace Research* 50, no. 3 (2013): 415–423.

Clausewitz, Carl von. *On War.* Translated and edited by Michael Howard and Peter Paret. Princeton, NJ: Princeton University Press, 1976.

Clemhout, Simone. "Typology of Nativistic Movements." *Man* 64 (January–February 1964): 14–15.

Collier, David, Jody LaPorte, and Jason Seawright. "Putting Typologies to Work: Concept Formation, Measurement, and Analytic Rigor." *Political Research Quarterly* 65, no. 1 (2012): 217–232.

———. "Typologies: Forming Concepts and Creating Categorical Variables." In *The Oxford Handbook of Political Methodology*, edited by Janet M. Box-Steffensmeier, Henry E. Brady, and David Collier. Oxford: Oxford University Press, 2008.

Collier, Paul. *Economic Causes of Civil Conflict and Their Implications for Policy.* Oxford, UK: Oxford University, April 2006.

Collier, Paul, and Anke Hoeffle. "Greed and Grievance in Civil War." *Oxford Economic Papers* 56 (2004): 563–595.

Collier, Paul, and Nicholas Sambanis, eds. *Understanding Civil War, Volume 1: Africa.* Washington, DC: The World Bank, 2005.

Connable, Ben, and Martin C. Libicki. *How Insurgencies End.* Santa Monica, CA: RAND Corporation, 2010.

Dassel, Kurt. "Civilians, Soldiers, and Strife: Domestic Sources of International Aggression." *International Security* 23, no. 1 (1998): 107–140.

Dassel, Kurt, and Eric Reinhardt. "Domestic Strife and the Initiation of Violence at Home and Abroad." *American Journal of Political Science* 43, no. 1 (1999): 56–85.

Davis, Gerald F., Doug McAdam, W. Richard Scott, and Mayer N. Zald, eds. *Social Movements and Organization Theory.* Cambridge: Cambridge University Press, 2005.

Davis, Paul K., and Kim Cragin (ed.), Darcy Noricks, Todd C. Helmus, Christopher Paul, Claude Berrebi, Brian A. Jackson, Gaga Gvineria, Michael Egner, and Benjamin Bahney. *Social Science for Counterterrorism: Putting the Pieces Together.* Santa Monica, CA: RAND Corporation, 2009.

Elman, Colin. "Explanatory Typologies in Qualitative Studies of International Politics." *International Organization* 59, no. 2 (2005): 293–326.

————. "Explanatory Typologies in Qualitative Analysis." Memo, Arizona State University, 2005.

Fearon, James D. "Iraq's Civil War." *Foreign Affairs*, March/April 2007. http://www.foreignaffairs.com/articles/62443/james-d-fearon/iraqs-civil-war.

————. "Why Do Some Civil Wars Last so Much Longer Than Others?" Working Paper, Stanford University, Stanford, CA, July 12, 2002. http://web.stanford.edu/group/ethnic/workingpapers/dur3.pdf.

Galtung, Johan. "Violence, Peace, and Peace Research." *Journal of Peace Research* 6, no. 3 (1969): 167–191.

Gamson, William, A. *Talking Politics.* New York: Cambridge University Press, 1992.

Gann, Lewis H. *Guerrillas in History.* Stanford, CA: Hoover Institution Press, 1971.

Ganor, Boaz. "Terrorist Organization Typologies and the Probability of a Boomerang Effect." *Studies in Conflict & Terrorism* 31, no. 4 (2008): 269–283.

Gerlach, Luther P. "The Structure of Social Movements: Environmental Activism and Its Opponents." In *Networks and Netwars: The Future of Terror, Crime, and Militancy,* edited by John Arquilla and David Ronfeldt, 289–310. Santa Monica, CA: RAND Corporation, 2001.

Geyer, Michael. "Resistance as an Ongoing Project: Visions of Order, Obligations to Strangers, Struggles for Civil Society." *Journal of Modern History* 64, suppl. (1992): S217–S241.

Glenn, Russell W. *Counterinsurgency in a Test Tube: Analyzing the Success of the Regional Assistance Mission to Solomon Islands (RAMSI).* Santa Monica, CA: RAND Corporation, 2007.

Goertz, Gary, and James Mahoney. *A Tale of Two Cultures: Qualitative and Quantitative Research in the Social Sciences.* Princeton, NJ: Princeton University Press, 2012.

Goffman, Erving. *Frame Analysis: An Essay on the Organization of Experience.* New York: Harper Colophon, 1974.

Gottlieb, Roger S. "The Concept of Resistance: Jewish Resistance during the Holocaust." *Social Theory and Practice* 9, no. 1 (1983): 31–49.

Granovetter, Mark. "Threshold Models of Collective Behavior." *American Journal of Sociology* 83, no. 6 (1978): 1420–1443.

Gros, Jean-Germain. "Towards a Taxonomy of Failed States in the New World Order: Decaying Somalia, Liberia, Rwanda and Haiti." *Third World Quarterly* 17, no. 3 (1996): 455–471.

Grossman, Herschell I. "A General Equilibrium Model of Insurrections." *The American Economic Review* 81, no. 4 (1991): 912–921.

Guevara, Ernesto Che. "The Cadres: Backbone of the Revolution." *Cuba Socialista*, September 1962. Published online in 2002 to the Ernesto Che Guevara Internet Archive, https://www.marxists.org/archive/guevara/1962/09/misc/x01.htm.

Gurr, Ted Robert. *Why Men Rebel.* Princeton, NJ: Princeton University Press, 1970.

Hahn, Erin, Jonathon Cosgrove, William Lauber, Guillermo Pinczuk, and David Danelo. *Developing a Typology of Resistance: A Structure to Understand the Phenomenon.* Fort Bragg, NC: United States Army Special Operations Command, 2015.

Hahn, Erin, and William Lauber. *Legal Implications of the Status of Persons in Resistance.* Fort Bragg, NC: United States Army Special Operations Command, 2014.

Heemsbergen, Luke, and Asaf Siniver. "New Routes to Power: Towards a Typology of Power Mediation." *Review of International Studies* 37, no. 3 (2011): 1169–1190.

Heidelberg Institute for International Conflict Research. *Conflict Barometer 2010: 19th Annual Conflict Analysis.* Heidelberg: Heidelberg Institute for International Conflict Research, 2010. http://hiik.de/en/konfliktbarometer/pdf/ConflictBarometer_2010.pdf.

———. *Conflict Barometer 2013.* Heidelberg: Heidelberg Institute for International Conflict Research, 2014. http://hiik.de/de/downloads/data/downloads_2013/ConflictBarometer2013.pdf.

Hironaka, Ann. *Neverending Wars: The International Community, Weak States, and the Perpetuation of Civil War.* Cambridge, MA: Harvard University Press, 2005.

Hoffman, Bruce. *Inside Terrorism.* New York: Columbia University Press, 1998.

Hollander, Jocelyn A., and Rachel L. Einwohner. "Conceptualizing Resistance." *Sociological Forum* 19, no. 4 (2004): 533–554.

Hopper, Rex D. "The Revolutionary Process: A Frame of Reference for the Study of Revolutionary Movements." *Social Forces* 28, no. 3 (March 1950): 270–279.

Huang, Hua-Lin. "Dragon Brothers and Tiger Sisters: A Conceptual Typology of Counter-Cultural Actors and Activities of American Chinatowns, China, Hong Kong, and Taiwan, 1912–2004." *Crime, Law & Social Change* 45, no. 1 (2006): 71–91.

Huntington, S. P. *Political Order in Changing Societies.* New Haven, CT: Yale University Press, 1968.

Jackson, Richard, and Helen Dexter. "The Social Construction of Organised Political Violence: An Analytical Framework." *Civil Wars* 16, no. 1 (2014): 1–23.

Jasper, James M. "A Strategic Approach to Collective Action: Looking for Agency in Social-Movement Choices." *Mobilization: An International Journal* 9, no. 1 (February 2004): 1–16.

Joes, Anthony James. *Resisting Rebellion: The History and Politics of Counterinsurgency.* Lexington, KY: University Press of Kentucky, 2004.

Johnston, Hank. "A Methodology for Frame Analysis: From Discourse to Cognitive Schemata." In *Social Movements and Culture*, edited by Hank Johnston and Bert Klandermans, 339–380. London: Routledge, 2003.

Judd, Nick. "Upgrading Civil Resistance? Gene Sharp's 'Methods,' Rewritten for 2012." *TechPresident* (blog), May 8, 2012. http://techpresident.com/news/22159/upgrading-civil-resistance-gene-sharps-methods-rewritten-2012.

Jütersonke, Oliver, Keith Krause, and Robert Muggah. "Guns in the City: Urban Landscapes of Armed Violence." In *Small Arms Survey 2007: Guns and the City*, edited by Eric G. Berman, Keith Krause, Emile LeBrun, and Glenn McDonald, 161–195. Cambridge: Cambridge University Press, 2007.

Kalyvas, Stathis N. "The Ontology of 'Political Violence': Action and Identity in Civil Wars." *Perspectives in Politics* 1, no. 3 (2003): 475–494.

Kalyvas, Stathis N., and Matthew Adam Kocher. "Ethnic Cleavages and Irregular War: Iraq and Vietnam." *Politics & Society* 35, no. 2 (2007): 183–223.

Klandermans, Bert. "Mobilization and Participation: Social-Psychological Expansions of Resource Mobilization Theory." *American Sociological Review* 49, no. 5 (October 1984): 583–600.

Krause, Lincoln B. "Playing for the Breaks: Insurgent Mistakes." *Parameters* 39, no. 3 (2009): 49–64.

Kuran, Timur. "Now Out of Never: The Element of Surprise in the East European Revolution of 1989." *World Politics* 44, no. 1 (1991): 7–48.

———. "Sparks and Prairie Fires: A Theory of Unanticipated Political Revolution." *Public Choice* 61, no. 1 (1989): 41–74.

Larson, Eric V., Derek Eaton, Brian Nichiporuk, and Thomas S. Szayna. *Assessing Irregular Warfare: A Framework for Intelligence Analysis.* Santa Monica, CA: RAND Corporation, January 2009.

Leonhard, Robert, ed. *Undergrounds in Insurgent, Revolutionary, and Resistance Warfare.* 2nd ed. Fort Bragg, NC: United States Army Special Operations Command, 2013.

Leslie, Glaister. *Confronting the Don: The Political Economy of Gang Violence in Jamaica.* Occasional Paper 26. Geneva, Small Arms Survey, 2010.

Lofland, John. "Charting Degrees of Movement Culture: Tasks of the Cultural Cartographer." In *Social Movements and Culture*, edited by Hank Johnston and Bert Klandermans, 294–338. London: Routledge, 2003.

Lund, Daulatram B. "Organizational Culture and Job Satisfaction." *Journal of Business & Industrial Marketing* 3, no. 3 (2003): 219–236.

Mabry, Tristan James. "Language and Conflict." *International Political Science Review* 32, no. 2 (2011): 189–207.

Machiavelli, Niccolò. *Art of War.* Translated and edited by Christopher Lynch. Chicago: University of Chicago Press, 2003.

Machlis, Gary E., and Thor Hanson. "Warfare Ecology." *BioScience* 58, no. 8 (2008): 729–736.

Mampilly, Zachariah Cherian. *Rebel Rulers: Insurgent Governance and Civilian Life during War.* Ithaca, NY: Cornell University Press, 2011.

Marrus, Michael R. "Jewish Resistance to the Holocaust." *Journal of Contemporary History* 30, no. 1 (1995): 83–110.

Marshall, Monty G., and Benjamin R. Cole. *Global Report 2014: Conflict, Governance, and State Fragility.* Vienna, VA: Center for Systemic Peace, 2014.

Marshall, Monty G., Ted Robert Gurr, and Barbara Harff. *PITF Problem Set Codebook.* Arlington, VA: Center for Global Policy, April 2009. http://globalpolicy.gmu.edu/political-instability-task-force-home/pitf-problem-set-codebook/.

Martin, Brian. "Online Onslaught: Internet-Based Methods for Attacking and Defending Citizens' Organisations." *First Monday* 17, no. 12 (2012). http://firstmonday.org/ojs/index.php/fm/article/view/4032/3379.

McAdam, Doug. *Political Process and the Development of Black Insurgency, 1930–1970.* Chicago: University of Chicago Press, 1982.

McCarthy, John D., and Mayer N. Zald. "Resource Mobilization and Social Movements: A Partial Theory." *American Journal of Sociology* 82, no. 6 (1977): 1212–1241.

McCormick, Gordon H., and Frank Giordano. "Things Come Together: Symbolic Violence and Guerrilla Mobilisation." *Third World Quarterly* 28, no. 2 (2007): 295–320.

McCormick, Gordon H., Steven B. Horton, and Lauren A. Harrison. "Things Fall Apart: The Endgame Dynamics of Internal Wars." *Third World Quarterly* 28, no. 2 (2007): 321–367.

Merari, Ariel. "A Classification of Terrorist Groups." *Terrorism: An International Journal* 1, no. 3/4 (1978): 331–346.

———. "Terrorism as a Strategy of Insurgency." *Terrorism and Political Violence* 5, no. 4 (1993): 213–251.

Miller, Benjamin, and Uri Resnick. "Conflict in the Balkans (1830–1913): Combining Levels of Analysis." *International Politics* 40, no. 3 (2003): 365–407.

Molnar, Andrew R., William A. Lybrand, Lorna Hahn, James L. Kirkman, and Peter B. Riddleberger. *Undergrounds in Insurgent, Revolutionary, and Resistance Warfare.* Washington, DC: Special Operations Research Office, 1963.

Naylor, R. T., "The Insurgent Economy: Black Market Operations of Guerrilla Organizations." *Crime, Law and Social Change* 20, no. 1 (1993): 13–51.

Negri, Antonio. *Insurgencies: Constituent Power and the Modern State.* Translated by Maurizia Boscagli. Minneapolis, MN: University of Minnesota Press, 2009.

Noakes, Lindsey. "Violent Divisions: Explaining Violence through Borders." Master's thesis, School of Oriental and African Studies, University of London, 2014.

O'Brien, Kevin J. "Rightful Resistance." *World Politics* 49, no. 1 (1996): 31–55.

Olson, Mancur. *The Logic of Collective Action: Public Goods and the Theory of Groups.* Cambridge, MA: Harvard University Press, 1965.

Olsson, Christian. " 'Legitimate Violence' in the Prose of Counterinsurgency: An Impossible Necessity?" *Alternatives: Global, Local, Political* 38, no. 2 (2013): 1–17.

O'Neill, Bard E. *Insurgency & Terrorism: From Revolution to Apocalypse.* 2nd ed., rev. Sterling: Potomac Books, 2005.

———. "Towards a Typology of Political Terrorism: The Palestinian Resistance Movement." *Journal of International Affairs* 32, no. 1 (1978): 17–42.

Pantucci, Raffaello. *Developments in Radicalisation and Political Violence: A Typology of Lone Wolves: Preliminary Analysis of Lone Islamist Terrorists.* London: The International Centre for the Study of Radicalisation and Political Violence, March 2011.

Paret, Peter, and John W. Shy. *Guerrillas in the 1960's.* Princeton Studies in World Politics; no. 1. New York: Praeger, 1962.

Paul, Christopher, Agnes Gereben Schaefer, and Colin P. Clarke. *The Challenge of Violent Drug-Trafficking Organizations: An Assessment of Mexican Security Based on Existing RAND Research on Urban Unrest, Insurgency, and Defense-Sector Reform.* Santa Monica, CA: RAND Corporation, 2011.

Paul, Christopher, Colin P. Clarke, and Chad C. Serena. *Mexico Is Not Colombia: Alternative Historical Analogies for Responding to the Challenge of Violent Drug-Trafficking Organizations.* Santa Monica, CA: RAND Corporation, 2014.

Pellerin, Cheryl. "Lynn: Cyberspace is the New Domain of Warfare." *DoD News*, October 18, 2010. http://archive.defense.gov/news/news-article.aspx?id=61310.

Perry, Walter L., and John Gordon. *Analytic Support to Intelligence in Counterinsurgencies*. Santa Monica, CA: RAND Corporation, 2008.

Petersen, Roger D. *Resistance and Rebellion Lessons from Eastern Europe.* Cambridge: Cambridge University Press, 2001.

———. *Western Intervention in the Balkans: The Strategic Use of Emotion in Conflict.* Cambridge: Cambridge University Press, 2011.

Peukert, Detlev. "Working-Class Resistance: Problems and Options." In *Contending with Hitler: Varieties of German Resistance to the Third Reich,* edited by David Clay Large, 35–48. New York: Cambridge University Press, 1991.

Pinker, Steven. *The Better Angels of Our Nature.* New York: Viking, 2011.

Poland, James M. *Understanding Terrorism: Groups, Strategies, and Responses.* 2nd ed. Upper Saddle River, NJ: Prentice Hall, 2005.

Polletta, Francesca. " 'Free Spaces' in Collective Action." *Theory and Society* 28 (1999): 1–38.

Porta, Donatella della. "Organizational Structures and Visions of Democracy in the Global Justice Movement: An Introduction." In *Democracy in Social Movements*, edited by Donatella della Porta, 16–43. Basingstoke: Palgrave Macmillan, 2009.

Price, David H. "Counterinsurgency and the M-VICO System: Human Relations Area Files and Anthropology's Dual-Use Legacy." *Anthropology Today* 28, no. 1 (2012): 16–20.

Ramsbotham, Oliver. "The Analysis of Protracted Social Conflict: A Tribute to Edward Azar." *Review of International Studies* 31, no. 1 (2005): 109–126.

Record, Jeffrey. *Beating Goliath: Why Insurgencies Win.* Washington, DC: Potomac Books, 2007.

Regan, Patrick M. *Civil Wars and Foreign Powers: Outside Intervention in Intrastate Conflict.* Ann Arbor, MI: University of Michigan Press, 2000.

Reuter, Christoph. "The Terror Strategist: Secret Files Reveal the Structure of Islamic State." *Spiegel Online*, April 18, 2015. http://www.spiegel.de/international/world/islamic-state-files-show-structure-of-islamist-terror-group-a-1029274.html.

Scheper-Hughes, Nancy. *Death without Weeping: The Violence of Everyday Life in Brazil.* Berkeley: University of California Press, 1992.

Schmid, Alex P. "Statistics on Terrorism: The Challenge of Measuring Trends in Global Terrorism." *Forum on Crime and Society* 4, nos. 1 and 2 (2004): 49–69.

Scobell, Andrew, and Brad Hammitt. "Goons, Gunmen, and Gendarmerie: Toward a Reconceptualization of Paramilitary Formations." *Journal of Political and Military Sociology* 26, no. 2 (1998): 213–227.

Segal, David R., and Molly Clever. "The Sociology of War." In *Oxford Bibliographies* in Sociology. http://www.oxfordbibliographies.com/view/document/obo-9780199756384/obo-9780199756384-0161.xml.

Shah, Shelly. "Social Movements: Meaning, Causes, Types, Revolution and Role," *Sociology Discussion* (online forum), 2013. http://www.sociologydiscussion.com/social-movements/social-movements-meaning-causes-types-revolution-and-role/2248.

Shapiro, Jacob N. *The Terrorist's Dilemma Managing Violent Covert Organizations.* Princeton, NJ: Princeton University Press, 2013.

Sharp, Gene. "The Meanings of Non-Violence: A Typology (Revised)." *The Journal of Conflict Resolution* 3, no. 1 (1959): 41–66.

———. *The Politics of Nonviolent Action.* Boston: Portor Sargent, 1973.

Shultz, Richard H. "Conceptualizing Political Terrorism: A Typology." *Journal of International Affairs* 32, no. 1 (1978): 7–15.

———. *Global Insurgency Strategy and the Salafi Jihad Movement.* INSS Occasional Paper 66. USAF Academy, CO: US Air Force Academy, April 2008.

Shy, John, and Thomas W. Collier, "Revolutionary War." In *Makers of Modern Strategy: From Machiavelli to the Nuclear Age,* edited by Peter Paret, Gordon Alexander Craig, and Felix Gilbert. Princeton, NJ: Princeton University Press, 1986.

Snow, David A., Louis A. Zurcher Jr., and Sheldon Ekland-Olson. "Social Networks and Social Movements: A Microstructural Approach to Differential Recruitment." *American Sociological Review* 45, no. 5 (October 1980): 787–801.

Snow, David A., and Remy Cross. "Radicalism within the Context of Social Movements: Processes and Types." *Journal of Strategic Security* 4, no. 4 (2011): 115–130.

Sobek, David. "Rallying around the *Podesta*: Testing Diversionary Theory across Time." *Journal of Peace Research* 44, no. 1 (2007): 29–45.

Staniland, Paul. *Networks of Rebellion: Explaining Insurgent Cohesion and Collapse*. Ithaca, NY: Cornell University Press, 2014.

———. "Organizing Insurgency: Networks, Resources, and Rebellion in South Asia." *International Security* 37, no. 1 (2012): 142–177.

———. "States, Insurgents, and Wartime Political Orders." *Perspectives on Politics* 10, no. 2 (June 2012): 243–264.

Sułkowski, Łukasz. "Typologies of organisational culture—multidimentional classifications." *Przedsiębiorczość i Zarządzanie* 14, no. 2 (2013): 173–182.

Sutton, Jonathan, Charles R. Butcher, and Isak Svensson. "Explaining Political Jiu-jitsu: Institution-Building and the Outcomes of Regime Violence against Unarmed Protests." *Journal of Peace Research* 51, no. 5 (2014): 559–573.

Symmons-Symonolewicz, Konstantin. "Nationalist Movements: An Attempt at a Comparative Typology." *Comparative Studies in Society and History* 7, no. 2 (1965): 221–230.

Tarrow, Sydney G. *Power in Movement: Social Movements and Contentious Politics*. Rev. and updated 3rd ed. Cambridge: Cambridge University Press, 2011.

Tilly, Charles. *Contentious Performances*. Cambridge: Cambridge University Press, 2008.

———. *Popular Contention in Great Britain, 1758–1834*. Cambridge, MA: Harvard University Press, 1995.

Treverton, Gregory F., and Seth G. Jones. *Measuring National Power*. Santa Monica, CA: RAND Corporation, 2005.

US Department of the Army. *Insurgencies and Countering Insurgencies*. Field Manual 3-24 (FM 3-24). Washington, DC: Headquarters of the Department of the Army, May 2014.

US Department of the Army. *Tactics in Counterinsurgency*. Field Manual 3-24.2(FM 3-24.2). Washington, DC: Headquarters of the Department of the Army, April 2, 2009.

US Government. *Guide to the Analysis of Insurgency*. Washington, DC: US Government, 2012.

US Joint Chiefs of Staff. *Department of Defense Dictionary of Military and Associated Terms.* Joint Publication 1-02 (JP 1-02). Washington, DC: US Joint Chiefs of Staff, January 15, 2015.

Voorhoeve, Joris. *From War to the Rule of Law: Peace Building after Violent Conflicts.* Amsterdam: University of Amsterdam Press, 2007.

Walder, Andrew G. "Political Sociology and Social Movements." *Annual Review of Sociology* 35 (2009): 393–412.

Watts, Stephen, Jason H. Campbell, Patrick B. Johnston, Sameer Lalwani, and Sarah H. Bana. *Countering Others' Insurgencies: Understanding U.S. Small-Footprint Interventions in Local Context.* Santa Monica, CA: RAND Corporation, 2014.

Weber, Max. "The Types of Legitimate Domination." In *Economy and Society: An Outline of Interpretive Sociology.* Edited by Guenther Roth and Claus Wittich, 212–216. Berkeley: University of California Press, 1978.

Weede, Erich. "On Political Violence and Its Avoidance." *Acta Politica* 39, no. 2 (July 2004): 152–178.

Weininger, Elliot B. "Pierre Bourdieu on Social Class and Symbolic Violence." In *Alternative Foundations of Class Analysis*, edited by Erik Olin Wright, 116–166. Madison, WI: Social Science Computing Cooperative, University of Wisconsin Madison, 2002.

Weinstein, Jeremy M. *Inside Rebellion: The Politics of Insurgent Violence.* Cambridge: Cambridge University Press, 2007.

Willems, Jurgen, and Marc Jegers. "Social Movement Structures in Relation to Goals and Forms of Action: An Exploratory Model." *Canadian Journal of Nonprofit & Social Economy Research (ANSESRJ)* 3, no. 2 (2012): 67–81.

Young, Aaron M. "Insurgency, Guerrilla Warfare and Terrorism: Conflict and its Application for the Future." *Global Security Studies* 2, no. 4 (2011): 65–76.

Yu, Zhiyuan, and Dingxin Zhao. "Differential Participation and the Nature of a Movement: A Study of the 1999 Anti-U.S. Beijing Student Demonstrations." *Social Forces* 84, no. 3 (2006): 1755–1777.

INDEX